Opening Process Equipment (OPE) Hazards

Opening Process Equipment (OPE) Hazards

Risk Recognition and Prevention

M. Darryl Yoes
Safety Consulting International, LLC
Jackson, LA, USA

WILEY

Published by John Wiley & Sons, Inc., Hoboken, New Jersey.

For general information on our other products and services or for technical support, please contact our Customer Care Department within the United States at (800) 762-2974, outside the United States at (317) 572-3993 or fax (317) 572-4002.

Wiley also publishes its books in a variety of electronic formats. Some content that appears in print may not be available in electronic formats. For more information about Wiley products, visit our web site at www.wiley.com.

Library of Congress Cataloging-in-Publication Data

Names: Yoes, M. Darryl author
Title: Opening process equipment (OPE) hazards : risk recognition and prevention / M. Darryl Yoes.
Description: Hoboken, N.J. : John Wiley & Sons, 2026. | Includes index.
Identifiers: LCCN 2026010073 | ISBN 9781394422241 cloth | ISBN 9781394422258 epub | ISBN 9781394422265 adobe pdf | ISBN 9781394422272
Subjects: LCSH: Industrial equipment—Safety measures | Industrial safety | Industrial equipment—Utilization | Manufacturing processes—Safety measures | LCGFT: Case studies
Classification: LCC TS191 .Y64 2026
LC record available at https://lccn.loc.gov/2026010073

Cover Design: Wiley
Cover Image: © madscinbca/stock.adobe.com

Table of Contents

- January 11, 2004; Huntsman Chemical, Port Neches
  Thermal decomposition of trapped liquid resulted in a fire while the piping was being prepared for maintenance work. The incident resulted in eight injuries, including two serious burn injuries.
- August 29, 2003; ExxonMobil Terminal, Nagoya, Japan
  A gasoline tank fire occurred in a tank being cleaned while hot work was in progress in an adjacent gasoline tank. Six fatalities, and one injury requiring hospitalization.

Series II:
Loss of Containment Due to the Failure to Properly Identify the Equipment for Opening, or the Workers Cutting or Opening the Wrong Equipment:

- February 1, 2025; Martinez Refining Company, Martinez, California
  A release and a significant fire occurred while opening the process equipment for maintenance. Local media reported that six workers were reported injured, four of whom were treated at a local hospital and released *(note this is very preliminary since this is a recent incident).*
- October 10, 2024; Pemex Refinery, Deer Park, Texas
  A major hydrogen sulfide release occurred while the blinds were being removed for maintenance.
  Two fatalities, 13 injuries, and the cities of Deer Park and Pasadena were ordered to shelter in place.
- October 17, 2016; BASF Chemical Plant, Ludwigshafen, Germany
  An explosion and fire occurred while a specialized pipeline construction company conducted preventative maintenance on a propylene line in a major pipe route. An explosion appeared on an adjacent ethylene line, and the fire quickly spread to other product and supply pipelines.
  Four people died, seven sustained serious injuries, and 22 others were injured in this incident.
- Date unknown (before 1990); Exxon Chemical Plant, Baton Rouge, Louisiana
  A major propane release occurred when pipefitters opened a 4″ flange on a propane manifold. The open flange was on the pressurized side of a valve on a propane line that had been fully prepared for opening. There was no ignition and no fire or explosion. This incident resulted in significant changes to line-breaking procedures.

Series III:
Loss of Containment Due to the Failure to Follow Procedures, or Noncompliance with Work Permit Requirements:

- July 2, 2020; Astron's Cape Town Refinery in Cape Town, South Africa

Preface

This book highlights key lessons learned and actions that can prevent incidents like the recent ones at the Pemex Refinery in Deer Park, Texas (October 2024, which resulted in two fatalities and a significant hydrogen sulfide release), and at the Martinez Refining Company in Martinez, California (February 2025, which caused a major fire and community disruption). It also covers lessons learned from numerous other incidents related to preparing equipment for Opening Process Equipment (OPE), or during the opening of process equipment for maintenance, inspection, or repairs. This book can save lives if these simple actions are adopted across our industry.

I was just finishing the first draft of my third Wiley Process Safety book, titled *Hydrogen Sulfide (H_2S) Hazards: Risk Recognition and Prevention*, when another tragic Process Safety incident occurred at the Pemex Refinery near Deer Park, Texas, resulting in two fatalities and injuring 13 others. This incident involved a massive release of highly concentrated hydrogen sulfide (H_2S) from an amine unit piping circuit, leading to a large shelter-in-place order for the communities of Deer Park and nearby Pasadena, Texas. As expected, this incident is discussed in the book *Hydrogen Sulfide (H_2S) Hazards: Risk Recognition and Prevention*. It is also directly related to OPE, the main focus of this book, and additional details will be available here.

In a report to local media, Pemex stated that the release occurred on October 10, 2024, while workers were installing blinds in a piping system and accidentally opened a flange on the wrong pipeline. In an update, the US Chemical Safety and Hazard Investigation Board (CSB) reported that the workers opened a flange about five feet away from the intended one. One of the workers opening the flange was killed, along with another worker who was working for a different company, less than 250 feet away and downwind from the release. The CSB almost immediately announced that it had dispatched a team to investigate the facility thoroughly. It may be some time before their report is available. However, it will come, and I expect the CSB report to be detailed, including photographs, interview results, and probably another informative process safety video. Nonetheless, we already know that this incident happened during the opening of process equipment or OPE, which can be very dangerous and has caused many injuries and loss of life, or forever changed lives.

This reminded me of several similar incidents that occurred during the opening of process equipment, although with various underlying and different causal factors. Being familiar with these tragic incidents in the oil and gas and petroleum refining and petrochemical industries, but not limited to our industries, I immediately started working on this book to help highlight and hopefully communicate more broadly what has happened and what can still happen today.

I was making good progress on the draft of this book. On February 1, 2025, about seven weeks after the Pemex incident, the Martinez Refining Company, located in Martinez, California, experienced a major fire in one of its process units. Fortunately, this incident resulted in no fatalities, although several workers were treated at a local hospital and later released. The refinery and local media reported that this incident occurred while workers were opening process equipment on a unit that had been taken offline just a few days before and was being prepared for planned maintenance. So far, the CSB has not indicated that it is investigating this incident; however, the California Contra Costa Health Services has begun an investigation, and hopefully, we will have an opportunity to understand more of the details leading to this incident, along with recommendations to help prevent another occurrence like this one.

Martinez Update as of May 6, 2025:

A draft report prepared by an external consultant and submitted to Contra Costa County Health Services concluded that the fire at the Martinez Refinery on February 1, 2025, was caused by contractors opening the wrong flange on a pipeline while attempting to install an isolation blind to isolate the piping circuit for maintenance. The report indicated that the piping circuit had been properly isolated, vented, drained, and prepared for opening. The contractors began opening the correct flange on the upstream side of the downstream isolation valve; however, they accidentally opened the downstream flange of the same valve instead. This side of the valve had not been prepared and was still pressurized. Working from a scaffold, the contractors released hot oil at approximately 630 °F (332 °C), which ignited about a minute after the release, likely due to autoignition. Fortunately, they managed to escape before ignition occurred. Although there were no serious injuries, this major fire caused extensive damage to the refinery and resulted in a prolonged shutdown of several units.

I am familiar with another similar incident at our site involving a contract crew as they were opening a four-inch diameter liquid propane process line to install an isolation blind. I don't remember the exact date, but I was there as a responder and remember the incident almost like yesterday. This pipeline was estimated to be nearly 10,000 feet (3,048 meters) long and was fully pressurized with liquid propane. More details about this incident are in this book, but the quick summary is that the contract workers picked up the work permit in the control room and headed to the job site. No joint job site visit or start-of-job discussion occurred in the field. The blind point was neither identified in the field nor on the process equipment drawing; there was no blind point tag, no paint marks, and no other identification provided. The pipefitters thought they knew exactly where the blind was to be installed, and they were on the correct pipeline. Unfortunately, they opened the flange on the pressurized side of an isolation valve. The 10,000-foot-long pipeline was pressurized by liquid propane. The opposite side of the isolation valve had been prepared for opening, but the flange that the pipefitters opened was not.

What followed was a massive propane vapor release when the flange was opened. Luckily, there were no ignition sources nearby, and there was no fire, no explosion, and no one was injured. However, as you can see, this was a very close call. The facility quickly acted to require joint job site visits in the field and a requirement for clear and distinctive identification for all flanges and access plates to be opened. Although this incident happened long ago, these requirements and others similar to these remain in place today.

These loss-of-containment incidents occur during OPE for many reasons or causal factors. A few of these causes are listed below, and we will discuss all of these details in this book, along with other contributing factors in our industry and other industries. Unfortunately, these incidents have happened many times before, and as you can see from the above discussion, they are still happening today. When they occur, they often lead to a significant fire or explosion, where workers have been badly burned or killed, and lives have been torn apart. These incidents can also cause chaos and disruption in our communities and severely damage the environment where we live and work.

The following is a list of potential causal factors that can lead to the loss of containment of flammable or toxic material while preparing equipment for opening or conducting OPE. All these causal factors have caused major loss-of-containment incidents and resulted in devastating fires or explosions. I believe this risk still exists at many facilities today. This list is not hypothetical and is not meant to be complete.

- Failure of the operations group to properly isolate a process pipeline or process equipment from other live equipment before handing it over to maintenance for line or equipment opening.
- Workers who open flanges and accidentally break a flange on a pipeline or process equipment that hasn't been prepared for opening and still contains pressurized process fluids or vapor.

- Worker familiarity with the job, but not with the specific task.
- Failure to properly purge, drain, or vent a pipeline, a blinding point, or process equipment that still contains trapped process fluids, vapors, or internal pressure before turning it over to maintenance personnel for opening.
- Failure to identify the risk of pyrophoric materials accumulating inside process vessels such as storage tanks, process fractionation columns, or other vessels before opening them to the atmosphere can lead to internal fires or explosions.
- Workers using a torch to open process equipment by burning the flange bolts to separate pipe flanges or vessel access plates. This can ignite trapped or unconfined, possibly pressurized, flammable liquids, vapors, or toxic materials, causing a fire, explosion, or toxic gas release.

Not only will we talk about the causal factors, but we'll also talk about something even more important: how these types of incidents can be avoided or prevented. In this book, we'll talk about not only how a loss of containment can happen during the execution of critical OPE tasks, but we'll also talk about why these incidents happen and, in a practical way, the actions or precautions that should always be in place to prevent them in the first place. These incidents can occur suddenly and without warning; again, lives can be changed forever in seconds. It is important to understand these precautions and ensure that actions are in place before the incident can occur.

You will also find a Chapter Quiz at the end of each chapter as a reminder of the important "learnings" in the chapter and an End-of-Book Quiz near the end of the book. The answers to each quiz are available in the appendix. I encourage you to do a quick review using this guide to help you ensure that you have picked up on the important points made.

It is believed that the information provided in the book will help improve safety performance in the petroleum and petrochemical industry and all other industries, especially for those involved in OPE. However, neither Safety Consulting International LLC, the author, M. Darryl Yoes, nor the company producing the documents contained herein warrants or represents, expressly or by implication, the correctness or accuracy of the content or the information presented here. This material is presented to improve operations' safety worldwide and is not intended to be used as operating guidance or procedures. Any use of the material contained herein is done with the user accepting legal liability or responsibility for the consequences of its use or misuse.

Safety Consulting International, LLC, the author, or any other company or person as outlined above, do not make any claim, representation, or warranty, expressed or implied, that acting in accordance with this book or its contents will produce any particular results with regard to the subject matter contained herein, or satisfy the requirements of any applicable federal, state or local laws and regulations. Nothing in this document constitutes technical advice. If such advice is required, it should be sought from an attorney, a qualified company, or a qualified individual.

No material known to be copyrighted or proprietary has been intentionally used in this document without the owner's express approval or permission. The author has made extraordinary efforts to contact all owners of images used in this book to secure consent to use photographs. All information presented is readily available in the public sector or represents the author's experience and over 50 years of collecting process safety information. The copyright owner approved most images; the author took those without attribution. There are several where, after an extensive search, the original could not be found; these are noted as such.

I want to be clear: the case studies are presented to share the lessons learned with others in the industry and, hopefully, to reduce future incidents. I am not placing blame and have no issues with the companies or individuals involved. It is evident from the Pemex and Martinez Refinery incidents that these situations are still occurring today, and people continue to sustain serious injuries or be killed in these events.

I hope you enjoy the book and have an opportunity to learn something about working with and around any process or mechanical tasks involving OPE and the safety aspects that go along with it. I strongly encourage you to share lessons learned with your peers. You may save a life by doing so.

Other Wiley Process Safety Books written by M. Darryl Yoes:

- *Process Operations Safety: The What, Why, and How Behind Safe Petrochemical Plant Operations.*
- *Hazards of Nitrogen and Other Inert Gases: How They Can Be Safely Managed.*
- *The Hazards of Hydrogen Sulfide (H_2S): How to Recognize the Deadly Hazards and Prevent Their Effects.*
- *Hazards of Opening Process Equipment and How They Can Be Successfully Mitigated.*

About the Author

M. Darryl Yoes is an experienced refinery supervisor and manager, having spent over 50 years in the refining industry. He began his career as a Process Apprentice at one of the largest US Gulf Coast refineries. After quickly advancing through several supervisory and management assignments, including first- and second-line supervisor positions in Process, Mechanical, and Technical, he advanced to the position of Risk Management Advisor. He then became the Safety, Health, and Environmental Manager of one of the largest petroleum refineries in the world.

He either led or participated in numerous process safety incident investigations, including those involving the tragic loss of life. He served as emergency coordinator and incident manager in the role of refinery superintendent and has extensive experience managing other emergency operations, including oil spill response operations. He has also led or participated in numerous process hazards analysis studies and led or participated in operations integrity and safety risk assessments at major US and European petrochemical sites. He has also served as the Safety Manager for major construction projects at US refineries.

Mr. Yoes continues the process of safety management consulting for refining and petrochemical plants and for construction safety management for major construction projects across the US, and is a course leader and facilitator for petrochemical plant Operations Safety Management training designed for refinery and petrochemical plant management, supervision, engineering, and plant operators. He has led comprehensive Process Safe Operations Training courses in essentially all regions of the world, including North and South America, Asia, and Europe, to share lessons learned from industry process safety incidents involving loss of containment of flammable or toxic materials leading to fire, explosions, and injuries or fatalities. The Safe Operations Training presents practical methods or work practices to help avoid a recurrence of these types of incidents.

Mr. Yoes is the author of three other Wiley books and online process safety resources titled *Process Operations Safety: The What, Why, and How Behind Safe Petrochemical Plant Operations*, *The Hazards of Nitrogen and Other Inert*

Gases: How They Can Be Safely Managed, and *Hydrogen Sulfide (H_2S) Hazards: Risk Recognition and Prevention*. They are available at bookstores, Amazon, and online. These books present a comprehensive overview of process safety in petroleum refineries, petrochemical plants, and other industries. They are valuable resources to operators, plant supervisors, engineers, and others responsible for process operations safety. To a significant extent, the content of the Safe Operations Training course is reflected in these books.

Mr. Yoes is also the founder and President of Safety Consulting International, LLC, which developed these helpful guides for preventing process safety incidents. He is employed by EcoScience Resource Group, LLC in Baton Rouge, a firm specializing in developing process operations training manuals, process operating and environmental procedures, and operator training guides for the petroleum refining and petrochemical industries.

1

What Is "Opening Process Equipment"?

Let's clarify what is NOT considered Opening Process Equipment (OPE). In most facilities, work performed by the operations group, also known as the process team, does not qualify as OPE. Tasks such as collecting samples, replacing various feed or product filters, opening built-in gauging hatches on storage tanks, tank cars, or tank trucks for sampling or gauging, and operating pig traps to insert or remove pigs are common process activities at most facilities. Venting or draining a product through a valve under controlled conditions, coke drum de-heading, or removing a plug from a bleeder valve are typically classified as process-related work and are not considered OPE.

This type of work is usually managed and supervised by the operators, and no work permit or special approval is generally required. Of course, there are always exceptions, and at some facilities, this might be classified as OPE, but in my experience, that is rare.

OPE involves performing mechanical work to open process drums, fractionator columns, or other vessels or equipment, exposing their interiors to the atmosphere. This exposure creates the risk of loss of containment, such as a spill or release of hydrocarbons, toxic chemicals, or pressurized products. It also allows outside air to enter, which can react with flammable or pyrophoric materials inside the equipment and potentially lead to a flammable or explosive atmosphere or ignition caused by accumulated pyrophoric materials. This work includes opening piping or vessel flanges, removing access plates, opening flanges to install or remove blinds, and opening small-diameter piping, instrument taps, tubing, and similar components.

Examples of OPE include opening flanges for blind insertion or removal, accessing plates on vessels for inspection or entry, removing exchanger heads, pumps, or compressor blinds, equipment removal, or reactor head removal, among other OPE-type tasks. These tasks require specialized training, tools, personal protective equipment, and other gear. The process equipment must be prepared for opening or made safe by the operations team. A comprehensive work permit must be approved by an authorized permit issuer from the process organization. Trained craftsmen perform the fieldwork.

"Opening process equipment" (OPE) also describes a type of failure that can cause loss of containment incidents, often leading to fires, explosions, and injuries to personnel when process equipment is opened for inspection, cleaning, or maintenance. This failure can happen with various types of process equipment, such as piping systems, storage tanks, or other process vessels like fractionation columns, pumps, compressors, or many other types of equipment.

This is sometimes called "breaking containment," but breaking containment actually refers to a loss of containment caused by various factors. For instance, operating a process vessel with the metal shell temperature below the minimum design metal temperature (MDMT) can cause metal embrittlement and vessel failure due to brittle fracture. Another example of breaking containment is not having proper procedures when performing a specific task or not following established procedures, which can lead to overfilling or overpressure of the vessel, ultimately resulting in vessel failure and a loss of containment.

The facility must clearly define the roles and responsibilities of all personnel involved in OPE work planning, equipment isolation, and preparation for opening. All personnel should be trained and qualified to safely perform their assigned tasks before assuming these responsibilities. A detailed procedure for removing flammable and toxic materials from process equipment should be developed and used as a training guide for operations personnel involved in OPE activities. It must be strictly followed during OPE execution. If circumstances require and the procedure cannot be followed exactly as written, a deviation process that requires preapproval should be used as a safe alternative. The deviation must be approved by at least one level higher in the organization before implementation.

The process organization typically ensures isolation from other energized or pressurized systems, which must be confirmed before the OPE work permit is issued for work to start. Energy can be defined as process pressure, the presence of hydrocarbons, or toxic materials. It can also come from electrical, spring, or hydraulic forces; all energy must be eliminated to the zero target. The goal of energy isolation is always to reach a "zero-energy present" state. If zero energy cannot be achieved, ongoing work should involve a risk assessment and mitigation measures to manage the remaining energy. The work plan, including mitigation strategies, should also undergo a higher-level review. For example, an identified leaking isolation valve should prompt a risk assessment to fully understand the additional hazards, how they are mitigated or controlled, and specific actions to ensure proper mitigation. OPE is a hazardous activity and requires a comprehensive joint site inspection to ensure clear communication and proper identification, addressing, or mitigation of hazards.

Frequent return visits to the job site should be expected to ensure that hazards are being identified on an ongoing basis, they are properly managed, and that job site hazards or conditions have not changed.

A Brief Look into Our Past – What Has Happened

The following list includes several examples of potential loss of containment incidents that have occurred. These incidents are not necessarily related to the opening of process equipment but are included here for informational purposes. I discuss these and other types of failures in detail in the first Wiley textbook, *Process Operations Safety: The What, Why, and How Behind Safe Petrochemical Plant Operations*. This list is not exhaustive and is not arranged in any particular order.

- Piping or equipment failure due to thermal expansion.
- Piping failure due to the installation of incorrect metallurgy in piping circuits (positive materials identification [PMI] or PMI failure).
- Undetected or uncontrolled exothermic reactions and thermal runaway reactions.
- Failure of an improperly supported and braced small piping connection.
- Ignition of flammable products due to failure to control static electricity.
- Sending a light or hot product to an atmospheric storage tank, resulting in the sinking of the floating roof or a tank boilover.
- Overfilling a storage tank due to a faulty level gauge, high-level alarm failure, or operator inattention.
- Failure to segregate incompatible chemicals resulting in an uncontrolled chemical reaction.
- Commissioning an unmanaged change resulting in equipment failure.
- Failure to properly return a safety-critical device to full online service after an equipment check or maintenance.
- Failure to follow operating procedures or exceeding the equipment operating envelope.
- Failure of a water or chemical injection point or failure of the piping downstream of an injection point.
- Failure of piping dead legs due to internal corrosion, erosion, or freezing water in the dead leg.
- Failure of a heater due to not following furnace procedures, especially during a restart following a furnace trip, or due to flow loss through the furnace tubes.
- Piping failure due to undetected corrosion under insulation.
- Catastrophic equipment failure while carrying out pneumatic pressure testing.

- Failure of equipment designed as a cold wall design (piping lined internally with refractory).
- Failure to follow proper water draw (drain) procedures on light-end vessels such as spheres.
- Ignition of combustible dust due to failure of standard housekeeping procedures.
- Failure to recognize or control the development of "popcorn polymer."
- Failure to recognize or control an uncontrolled ethylene decomposition reaction.

All these failures can lead to catastrophic consequences, so we need systems and processes to prevent them. However, this book will specifically focus on the loss of containment that can occur during equipment preparation, isolation, or when opening process vessels and other equipment. Unfortunately, this can and does happen, and when it does, the result is often a major fire or explosion. Many people have been killed while preparing equipment or OPE for inspection or maintenance.

A common safety incident during OPE processes occurs when workers open flanges to install a blind, flat plate with gaskets on each side to isolate equipment for access or entry. The incident can happen if pressure or temperature remains in the system being opened due to valve leakage, failure to isolate the system from other pressurized equipment, or failure to verify that pressure has been relieved before opening the flange. When this happens, workers opening the flange might be sprayed with process fluid, which can be hot or toxic, and it could also ignite a flash fire that engulfs the workers, causing severe injury or death.

This also occurs when workers navigate along a pipe rack, opening flanges on a single pipeline that runs parallel and is mixed with many other piping systems. Workers can, and in some cases have, mistakenly access the wrong line and open a flange on a pipeline that remains pressurized. When this happens, the pipeline they open may contain highly flammable or toxic substances. Such incidents have occurred, and the outcomes can be tragic. According to preliminary reports from the CSB, this situation might be similar to what happened in the Pemex incident on October 10, 2024.

Loss of containment also occurs when workers open storage tanks or process fractionation columns before completely removing the process fluid or vapors from the vessel. Another example that has led to tank fires or the collapse of process fractionators is opening the vessel without recognizing the presence of pyrophoric material inside. This material is usually an iron sulfide scale that accumulates in the storage tank or fractionation column during recent onstream processing. It is pyrophoric and can ignite rapidly once air enters the process vessel. The main hazard is that once ignited, other flammables or combustibles can easily catch fire from the pyrophoric fire. This includes metal fractionation internal components, such as demister screens or metal alloy fractionation trays.

Process fractionators have collapsed and toppled due to rapid overheating of the tower shell and support structure, along with an internal fire caused by the ignition of pyrophoric material. The pyrophoric material often ignited other flammable or combustible residual materials. As mentioned above, the metal fractionation column internals, such as demister screens or random or structured metal packing material, are easily ignited by pyrophoric substances like iron sulfide scale.

Although not currently mandated by the OSHA Process Safety Standard (OSHA 29 CFR 1910.119), I have long recommended that sites develop and use dedicated work permits for all OPE work activities. I view the dedicated work permit as a way to help ensure additional focus on this highly hazardous task and to facilitate a checklist that ensures the critical aspects of OPE are incorporated into the work plan.

A separate OPE work permit with a distinctive color clearly emphasizes this type of work and sets it apart from more routine cold work tasks. I understand that the current cold work permit form usually includes specific checkboxes for isolation, depressurization, and electrical lockout, which are also relevant for OPE work. However, I believe that having a separate form would allow for more detailed checkboxes and highlight other critical criteria specific to OPE, such as the following (please note these are only examples and not an exhaustive list):

- The requirement for a process standby to be present at each first break into each piping circuit and/or each process vessel.
- Requiring a provision to safely collect any potential spillage for each flange or access plate that is opened.
- Vents and drains are verified as free, clear, and in a zero-energy state before a work permit is issued. If a vent or drain is fouled or plugged, a proper bleeder unplugging device will be used to help verify that the bleeder is clear before proceeding.
- If H_2S is suspected or identified via the preopening gas test, breathing air (either a self-contained breathing apparatus (SCBA) or a hose line supplied) is required.
- The location of blinds and equipment openings is identified (tagged in the field) before the OPE permit is issued. A weather and tamper-resistant tag should be placed on each point of line or vessel opening. Each isolation tag should be numbered in a manner consistent with the line opening procedure and blind list.
- The sequence of blind installation and removal considers the potential of trapped liquids or pressure. A vented blind should be used if the potential creation of a trapped blind is identified.

- If zero energy cannot be confirmed, no flange bolts are removed. It is acceptable to remove one bolt at a time and replace it with a new one (please see "Hot Bolting" in Chapter 2).
- The potential presence of pyrophoric material should be considered and addressed before OPE. For vessels with the potential for pyrophoric materials to be present, providing a charged fire water hose at each vessel access plate or manway before the access plate is opened is good practice. The charged fire hose can periodically wet down the vessel interior, thereby preventing ignition of the pyrophoric material.
- Always confirm that there is no hot work in progress within a safe distance of the equipment opening.
- All OPE work will cease immediately, and flanges will be retightened if any leakage or spillage occurs.
- If the conditions of this permit cannot be met, ensure a risk assessment is conducted, and a deviation is approved before the OPE permit is issued.

As stated above, these loss-of-containment incidents often occur when breaking flanges or opening equipment for blinding or equipment isolation. They can occur during online maintenance, startup operations, or during shutdown operations. Table 1.1 provides several examples of operations typically classified as OPE.

Table 1.1 Examples of operations or maintenance tasks that are typically classified as opening process equipment (OPE) or breaking containment.

- Isolating or removing a process pump for maintenance activities, such as routine overhaul, bearing, or seal replacement.
- The isolation of a filter or coalescer for cleaning or desiccant replacement, or a salt drier for salt replacement.
- Isolating a piping circuit for maintenance to replace a gasket or an isolation valve.
- Isolating process instrumentation for instrument maintenance or replacement.
- Isolating a section of process piping for maintenance, pipe modifications, or piping replacement.
- Isolation of process equipment, such as a fractionation column, a process vessel, or a storage tank for internal inspection, cleaning, or maintenance.
- Isolation of a piping circuit or process vessel to allow the installation of isolation blinds or the removal of a blind flange on the end of the piping.
- Isolation of a section of piping or process equipment to allow installation of new equipment.
- Isolation of a pressure relief valve for valve replacement or shop overhaul, inspection, or maintenance.
- Isolation of piping circuits to facilitate the unplugging of piping or valves.
- Isolation of a control valve for maintenance or replacement.

Table 1.2 illustrates several examples of loss of containment incidents and details the unit's operational stage at the time of the incident. Chapter 11, the What Can Happen will cover these in more detail, including additional loss of containment incidents involving OPE.

Table 1.2 Table illustrating loss of equipment incidents which occurred during equipment preparation or OPE for cleaning or maintenance (not a complete list).

Tosco Avon Refinery February 23, 1999	Crude Unit – Fire during maintenance work to replace a section of the overhead naphtha line 4 people were killed and 1 injured	The Crude Unit was in service when this incident occurred
ExxonMobil, Baton Rouge Refinery November 22, 2016	Alkylation Unit – In adequate isolation and knowledge of valve operation. Flammable vapor was released during unplanned maintenance around a pump. Although there was no explosion, the release ignited and caused a large fire	The Alkylation Unit was in service when this incident occurred
ExxonMobil, Beaumont Refinery April 17, 2013	Catalytic Hydro Deconversion (CHD) Unit Exchanger Fire while preparing for unit turnaround (12 injured, 2 fatalities)	The CHD Unit was down for planned turnaround. Exchangers were being opened for bundle removal
Giant Industries, Ciniza refinery, Jamestown, NM. April 8, 2004	Alkylation Unit – A loss of containment of hot alkylate and a fire occurred while mechanics were in the process of opening the pump flanges. 6 injured, and nearby facilities were evacuated, refinery equipment and support structures were damaged	Alkylation Unit – The hydrofluoric acid alkylation unit was in service. The spare alkylate product pump was undergoing maintenance for seal replacement
Chevron Pembroke (UK) Amine regeneration unit explosion June 2, 2011	Amine Unit – An explosion resulted from ignition of a flammable atmosphere in an amine storage tank during a routine tank cleaning operation 4 Fatalities and 1 life changing injury	The storage tank at the amine unit was undergoing routine cleaning in preparation for maintenance at the time

(Continued)

Table 1.2 (Continued)

Delaware City Refining Company, Delaware City, DE. November 29, 2015	Flash fire while preparing process equipment for Maintenance 1 employee injured (severe burns)	Sulfuric Acid Alkylation Unit – While draining what was believed to be steam condensate to a sewer, a flash fire occurred. The line had been isolated with a single isolation valve which leaked light hydrocarbons from a pressurized source into the line being drained
Phillips Petroleum Pasadena, TX. October 23, 1989	Release of ethylene and isobutane from reactor line during maintenance work resulting in massive vapor cloud explosion 23 fatalities, $1.6 billion in property damage and business losses	The chemical plant was in service at the time of the explosion. Maintenance was in progress to unplug plastic from a reactor settling leg when the release occurred

Chapter 11 is an expanded version of this table. It provides a broader view of examples of loss of containment incidents during OPE activities (equipment preparation, isolation, blind setting or removal, or equipment opening for cleaning, entry, or maintenance).

The State of California has developed a state regulation to effect oversight of the California petroleum refineries, which is recognized as state code § 5189.1, or the "Process Safety Management for Petroleum Refineries." This code requires the employer to develop, implement, and maintain safe work practices to prevent or control hazards during specific activities, including OPE or piping, that should be controlled. The paragraph from the California regulation is available below and can be downloaded from the link in the Additional References section of this Chapter.

> The employer shall develop, implement and maintain safe work practices to prevent or control hazards during specific activities, such as: opening process equipment or piping; tasks requiring lock-out/tag-out procedures; confined space entry; handling, controlling and stopping leaks, spills, releases and discharges; and control over entry into hazardous work areas by maintenance, contractor, laboratory or other support personnel. Safe work practices shall apply to employees and employees of contractors.

End-of-Chapter Quiz

1 Which of the following are NOT considered opening process equipment? Please choose all that apply.

a Opening flanges on a pipeline to install isolation blinds.
b Opening access plates on a storage tank to access the tank for cleaning and internal inspection.
c Opening a suction strainer on a product pump to clean the strainer basket.
d Opening an access plate on a chemical railcar to gauge the level in the tank.
e Opening the access plate on a product fractionator column for entry.
f Draining water from a crude oil storage tank before lining up the tank to the crude unit.
g All of the above.

2 Opening equipment to the atmosphere creates the possibility of a loss of ______________, such as a ________ or __________ of hydrocarbons, toxic chemicals, or a pressurized product. It also allows outside ________ to enter the equipment, which can react with flammable or ______________ material.

3 The facility is responsible for clearly defining the roles and responsibilities of all personnel involved in ________ work planning and equipment isolation and preparation for opening. All personnel must be __________ and _________ to safely perform and execute their assigned _________ before taking on this responsibility.

4 What are several potential advantages of a dedicated work permit for authorizing opening process equipment? Please select all that apply.

a A unique work permit dedicated to OPE would add additional emphasis to the hazards of OPE.
b A unique work permit may help facilitate a checklist of potential OPE concerns to be addressed.
c A unique work permit would facilitate more in-depth tick boxes, placing additional emphasis on other critical criteria that are more specific to OPE.
d A unique work permit would set OPE work apart from other more routine cold work tasks, adding emphasis to OPE work.
e All of the above.
f None of the above.

5 Based on industry data, at what point do many of the loss-of-containment incidents often occur?
Please select all that apply.
  a OPE incidents have occurred during online maintenance operations.
  b OPE incidents have occurred during startup operations.
  c OPE incidents have occurred during equipment blinding for isolation.
  d OPE incidents have occurred during shutdown operations.
  e All of the above.
  f None of the above.

6 If a leak or release occurs while conducting OPE tasks for equipment isolation or entry, what should take place immediately?

7 Vents and drains are verified as free and clear and in a ________-________ state before a ________ ________ is issued.
If a ________ or ________ is fouled or plugged, a proper ________ ________ device will be used to help verify that the bleeder is clear before proceeding.

8 What type of tag should be placed on all points of equipment blinding before the work permit for blinding is issued? When should this be installed at each blind point?

9 What special PPE is typically required if H_2S is suspected, or if it is identified during the preopening gas test?

10 Why is it important to consider the potential of creating a "trapped blind" when installing blinds for equipment isolation?

11 How can the creation of a trapped blind be prevented?

12 What is a special consideration when opening flanges on piping circuits or process equipment?
Please select the most correct answer.
  a Consider the potential for creating a trapped blind.
  b Ensure the correct bolts are being used.
  c Ensure that all bolts are properly lubricated.
  d Ensure the provision for collecting any trapped liquids.
  e All of the above.
  f None of the above.

Additional References

American Institute of Chemical Engineers (AICHE), Line Opening – Need / Call to Action, https://www.aiche.org/ccps/resources/tools/safe-work-practices/line-opening/need-call-action.

American Institute of Chemical Engineers, Center for Chemical Process Safety, Line Opening – Fundamental Intent, https://www.aiche.org/ccps/resources/tools/safe-work-practices/line-opening/fundamental-intent.

Beaumont Enterprise, ExxonMobil Agrees to Penalties 2013 Fatal Fire at Beaumont Refinery, https://www.beaumontenterprise.com/news/article/Exxon-Mobil-agrees-to-penalties-in-2013-fatal-13668898.php.

primeessay.org, Blinding Process in Refineries, https://primeessay.org/samples/research/blinding-process-in-refineries.

State of California; State Regulation – State Code 5189.1, Process Safety Management for Petroleum Refineries, https://www.dir.ca.gov/title8/5189_1.html.

US Chemical Safety and Hazard Investigation Board (CSB), ExxonMobil Baton Rouge Refinery Chemical Release and Fire, https://www.csb.gov/exxonmobil-baton-rouge-refinery-chemical-release-and-fire/.

US Chemical Safety and Hazard Investigation Board (CSB), Giant Industries Refinery Explosions and Fire, https://www.csb.gov/giant-industries-refinery-explosions-and-fire/.

US Chemical Safety and Hazard Investigation Board (CSB), Removal Of Hazardous Material From Piping Systems, https://www.csb.gov/assets/1/20/pipeopeningbulletin.pdf?13882.

US Occupational Safety and Health Administration (OSHA), 29 CFR 1910.119 – Process Safety Management of Highly Hazardous Chemicals, https://www.osha.gov/laws-regs/regulations/standardnumber/1910/1910.119.

US Occupational Safety and Health Administration (OSHA), OSHA Inspection Detail – Exxon Mobil Corporation, https://www.osha.gov/ords/imis/establishment.inspection_detail?id=901522.015.

2

Equipment Preparation for Opening – Making It Safe

Occasionally, equipment, such as piping circuits, process vessels, pumps, compressors, etc., at petroleum refining or petrochemical plants must be taken offline for cleaning, inspection, or to accommodate expansion projects. Generally, this equipment may have been operating in highly hazardous hydrocarbon service for several years or with highly toxic or thermally reactive chemicals. The challenge for operations is that this equipment must be safely and environmentally prepared for opening for access and entry by clearing the hazardous material, allowing air into the equipment without allowing a flammable atmosphere, and making the equipment safe for human entry. This means that a wide range of equipment must be prepared, including, to name a few, storage tanks, reactors, process piping, heat exchangers, fractionation columns, mixers, chillers, pumps, compressors, etc.

The preparation and opening of refining or petrochemical plant piping and other associated equipment can be very hazardous and require specialized procedures, well-trained personnel, and specialized personal protective equipment (PPE). Each US facility is required by the OSHA Process Safety Management Standard (29 CFR 19.10.119) to have the required process safety information and procedures required to develop an effective and safe unit shutdown and preparation procedure. This includes the process and instrument drawings, materials of construction, the shutdown procedure, the process safety information, the properties of the chemicals used in the process, the information on the unit's safety systems, etc. Using this information and the other OSHA PSM unit-specific information, the facility should develop a detailed shutdown and unit preparation procedure to be used by the operations organization to prepare the unit for clearing of hazards and for opening the equipment for access, cleaning, inspection, and repairs as required.

The procedure and drawings should be used in conjunction with a detailed walk-down of the isolation and clearing procedure to identify low-point drains and high-point vents, areas of trapped liquids, locations of steam tracing, blinding points, other equipment isolation points, and other similar sensitive areas that should be included in the procedures.

The following is a discussion of the general guidelines that may be helpful during equipment isolation, decontamination, blinding, and preparation for opening.

It is difficult to describe the generic isolation and clearing process due to the significant differences in equipment design, construction material, and operations characteristics. There are as many ways to clear a unit as there are different processes. I will describe in a very generic way the clearing operation for a crude unit, a unit without catalyst, and for comparison, a hydrotreater, a unit with reactors and catalyst.

To clear a crude unit of crude oil and other hydrocarbon compounds, the first step is to reduce the feed rate to a low but stable rate and divert the side streams away from the on-specification dispositions to prevent sending off-specification products. The incoming feed is then switched from crude oil to a much lighter feed, typically a medium gas oil feed, allowing the flushing of the desalters, fractionation columns, furnaces, and all other unit equipment. At this point, the furnace outlet temperatures are reduced to lower the boiling point in the fractionation section and allow the unit to run another few hours to flush the heavier products from the system fully.

We then gradually lower the feed rate and boiling point (furnace outlet temperatures) and slowly bring the feed rate to zero. At this point, the furnaces are tripped off by closing the fuel gas control valves, and isolation blinds are quickly installed in the fuel gas lines. This prevents unburned fuel from leaking into the furnace firebox, creating a flammable atmosphere inside the firebox. At this point, water is introduced into the feed train to displace the gas oil from the feed train and return it to storage. Once the feed train is water-filled, the water flow is stopped, and all low-point drains are opened on the feed train to drain the free water. The water is drained to the process sewer system and to the wastewater treatment plant, where the free oil is separated from the water. Once the free water is removed, steam is introduced into the feed train, furnaces, and fractionator columns to begin steaming the equipment to the atmosphere through the high-point vents and low-point drains. After steaming, the isolation blinds are set in preparation for chemical cleaning the feed train and/or fractionation columns in preparation for opening.

A hydrotreater is handled very differently due to catalysts in the reactors and a significant amount of stainless steel installed in the process piping, vessels, and exchangers. The primary difference is that we do not allow moisture into the reactor catalyst beds to prevent damaging the catalyst. Instead, we purge the reactors with nitrogen to remove the hydrocarbons. This process is called concertina purging or pressurizing, which rapidly pressurizes and depressurizes the reactors several times. After clearing the reactors, they are blocked under nitrogen pressure, and the reactors are gas tested to ensure air has been removed. Any water allowed into the feed train is pretested for chlorides before use to prevent stress corrosion cracking of the stainless-steel piping and process vessels.

From here, we proceed with the isolation blinds to segregate the unit into discrete sections, with the isolation blinds placed as close to each vessel or system as possible. We prefer that the isolation blind be placed directly into the first flange of each vessel or piping system. This effectively eliminates branch connections or tie-ins to other systems or parts of the unit and means less opportunity for contamination or cross-connection to the isolated vessel.

Initial Decontamination (Operations Team)

Note: The method used will differ depending on the process involved, the materials used for construction, and the contaminants to remove. This section covers the initial decontamination of process vessels and other equipment to remove hydrocarbons and other contaminants and the primary methods used.

- Draining from low-point drains;
- Flushing with a clean process stream such as light gas oil (for heavier products);
- Flushing with gas (hydrogen, natural gas, or fuel gas) to remove hydrogen sulfide from sour streams;
- Purging with nitrogen (concertina purging is most effective) (huff and puff or pressurize/depressurize);
- Steaming to a closed system or the atmosphere;
- Water washing; or
- Chemical cleaning or chemical neutralization.

Note: Some protective equipment may not be in service; therefore, when purging, flushing, or steaming equipment, always ensure that the vessel is operated below the Maximum Allowable Working Pressure (MAWP) (to prevent overpressure) and above the Minimum Design Metal Temperature (MDMT) (to prevent brittle fracture).

Additional Guidance on Steaming of Process Equipment

High-pressure steam is one of the most common methods of decontaminating process equipment and preparing it for opening. There are some good general safety guidelines to use when using steam for this purpose, as follows:

- Always be aware of the potential for steam burns or burns from hot condensate. All personnel involved in this activity should wear standard PPE such as gloves, face shields, rubber boots, etc.
- Steam hoses should be identified as suitable for high-pressure steam use and properly rated for the temperature and pressure.

- Steam hoses should also be equipped with a whip check, a safety device installed between two steam hoses or between the manifold and hose to prevent personnel injury from a flailing hose.
- Quickly address any water hammering to prevent hose or piping failure.
- Steam should not be injected into low-point drains as this interferes with condensate draining.
- Ensure condensate is being continuously drained from all low points, including dead ends.
- Ensure that a slight positive pressure is maintained on the equipment to keep the temperature hot.
- Ensure system pressure is monitored continuously and is maintained below the MAWP.
- Continue steaming until a gas test proves the system is gas-free.

Safety Considerations When Purging with Nitrogen

Nitrogen safety hazards are covered in detail in my second Wiley textbook titled *Hazards of Nitrogen and Other Inert Gases: How They Can Be Safely Managed.* They are included here due to the significance of these hazards and the number of tragic incidents that have occurred due to nitrogen and other inert gases:

> There are three primary hazards associated with nitrogen. In either form, as a liquid or gas, the deadly characteristic is that nitrogen will quickly displace oxygen from the environment. This has resulted in many incidents of accidental asphyxiation of workers. Please remember, it is not the nitrogen that kills; people die due to oxygen deprivation when the nitrogen has displaced the oxygen. See the case studies in the textbook mentioned above and its Appendix for a discussion of nitrogen-related incidents.
>
> Nitrogen is a frequent cause of fatalities and the primary subject of the referenced textbook. What follows is a summary; please refer to the Safety Data Sheet (SDS) or sometimes referred to as the Material Safety Data Sheet (MSDS) for additional information. The following contains a general description of nitrogen from the US Environmental Protection Agency (EPA) Cameo Chemicals Database:
>
> Nitrogen is a colorless and odorless gas. It is non-combustible and non-toxic, making up the majority of the atmosphere, but it cannot support life by itself. Nitrogen may cause asphyxiation by displacing air containing oxygen, which is necessary to sustain life. Under prolonged exposure to fire or heat, containers may rupture violently and explode.
>
> Health Hazards of Nitrogen from the EPA Cameo Chemicals Database and the Safety Data Sheet:

Nitrogen may cause dizziness or asphyxiation without warning. The victim may not be aware that they are being overcome. Cold vapors from liquefied nitrogen gas are initially heavier than air and spread along the ground.

High concentrations of nitrogen may cause asphyxiation. Symptoms may include loss of mobility and loss of consciousness. The victim may not be aware of asphyxiation. To protect yourself, wear a self-contained breathing apparatus (SCBA) when responding to someone who has been overcome. Remove the victim to an uncontaminated area and keep the victim warm and rested. Call a doctor. If breathing ceases, apply artificial respiration.

If you are working near a vessel that is being purged with nitrogen, appropriate warnings and barricades must be in place to prevent personnel from entering areas that may have an oxygen deficiency. Barricades must be designed to keep personnel from entering these areas, including all areas where nitrogen is vented into the atmosphere.

Personnel should be kept away from platforms near open access plates or process vessels with nitrogen venting. Those working near nitrogen vents must wear either a hose line-supplied breathing air respirator or a self-contained breathing apparatus (SCBA). Additionally, personnel wearing SCBA should continuously monitor others working in or near areas where nitrogen purge is active. No one should be on the platform or near the vessel opening where nitrogen is venting without breathing air (hose line or SCBA).

Warning: When in an oxygen-deficient atmosphere, a cartridge or chemical respirator is insufficient for breathing protection. The respirator must be connected to a certified breathing air supply line or have self-contained breathing air.

Do I have to be inside a confined space to be overcome by nitrogen?

> NO, you do not! There have been several cases where workers were outside a vessel being purged with nitrogen and collapsed due to oxygen deprivation, falling into the vessel and dying, or they were overcome, fell, and died from the fall.

Additional Guidance on Preparation for Pyrophoric Materials

There have been many occasions when storage tanks or process vessels have been opened only to ignite inside the vessel due to the unexpected presence of pyrophoric materials. Of course, when this happens, it is a total surprise, and it can result in the total collapse of fractionator columns or major fires in petroleum storage tanks.

We are most susceptible to this when the process service is a sour service with a high concentration of hydrogen sulfide (H_2S). The pyrophoric material, typically iron sulfide scale, lies dormant in the vessel(s) for years while the vessel is in service. However, air is allowed inside when the vessel is removed from service and opened to the atmosphere, causing the vessel to dry internally. This results in an exothermic reaction of the pyrophoric material and ignition of other flammable or combustible material inside the vessel. This has resulted in igniting metal demister screens and internal tower packing materials, which burn extremely hot and are very difficult to extinguish.

There is a lot more detail on this topic in the Wiley textbook by this author, *Process Operations Safety: The What, Why, and How Behind Safe Petrochemical Plant Operations*. The following presents several general precautions for pyrophoric materials during OPE operations:

- Always be aware of the potential for pyrophoric materials anytime equipment in sour service (H_2S) is being opened (such as storage tanks, fractionation columns, gas compressors, etc.).
- When opening equipment, place a charged fire hose strategically located at each manway being opened.
- Periodically spray water into the vessel manway and never allow the vessel interior to dry; otherwise, the pyrophoric material may ignite and ignite other flammable or combustible material. Water flooding of the column is not required; periodic wetting is needed to prevent the vessel from drying.
- During work preplanning, using a reputable chemical vendor, consideration should be given to pyrophoric neutralization of the vessel interior using a weak solution of potassium permanganate. A reputable chemical vendor can help with this.
- In the work plan, we should expect to remove the demister screen and internal packing material from all areas where hot work is planned. Otherwise, they may ignite and create an internal fire that is extremely difficult to extinguish.

Selecting the Equipment Isolation Points for Blinding (Operations Team)

When selecting the blind points, the objective is to isolate the equipment near the individual process vessels or other equipment to be blinded. Typically, the isolation blind is placed in the first flange off the vessel. This helps minimize the interconnections with other equipment and the opportunity to miss a connection.

It also helps eliminate the creation of dead legs or dead ends that may trap liquids or contaminate a vessel with hydrocarbons or toxics, possibly creating issues with a satisfactory gas test.

When identifying blind points, be careful about creating a "trapped blind," a blind that is pressurized on one side without a bleeder to vent or drain the trapped liquid or pressure. A vented or bleeder blind can be used if the piping geometry will create what will become a trapped blind.

When installing blinds, the craftsmen should install new gaskets on the side facing the process. Care must also be taken to protect the new gasket and prevent damage caused by flange wedges or other tools used in the blinding operation.

Considerations for Blinding Fuel Supply to Furnaces, Heaters, or Boilers (Operations with Craftsmen)

Furnaces, heaters, and boilers are typically fired by fuel gas, natural gas, or fuel oil and represent unique fire and explosion hazards. They are a confined space and have fuel connections directly into the firebox. When shutting these devices down, the fuel supply is blocked as close to the firebox as possible, typically at the fuel gas control valve station; however, these valves can leak through and result in a flammable mixture inside the firebox. This has resulted in devastating explosions, so these blinds must be installed as quickly as possible after the furnace or boiler is taken offline.

- To prevent the possibility of an explosion, the isolation blinds in the main fuel piping circuits on furnaces, heaters, and boilers should be installed as soon as the equipment is shut down.
- In the event of delays in blind installation, mitigation may include leaving the pilots in service, which will consume any unburned fuel, thereby preventing the creation of a flammable or explosive atmosphere in the firebox.
- If the furnace is equipped with induced draft or forced draft fans, another alternative is to leave the fans in service, which will ensure that any excess fuel is purged from the firebox.

Considerations for Blinding Flare Headers (Operations with Craftsmen)

Flare systems are the last line of defense to protect process equipment and systems from overpressure and must remain in service until all other equipment is down and safely vented. Generally, this means that the flare system is the last to be brought offline and only after all other operating units are safely down and secure.

Flare systems and flare stacks also represent unique hazards when shutting equipment down and preparing for a restart. Flare systems typically include very large diameter piping and are generally not designed to support the weight of water. These lines must be cleared by steaming or purging with nitrogen or other inert gas. Flare systems also generally handle high concentrations of hydrogen sulfide (H_2S), which present other safety risks to workers.

Also, the flare stack will create a natural draft and can pull significant amounts of air into it. If this occurs before the flare equipment is gas-free, a flammable mixture can easily be created.

Typical precautions while blinding or de-blinding flare systems or flare stacks:

- If steaming is used for clearing flare systems, precautions must be taken to ensure that condensates do not accumulate in the large piping circuits; otherwise, the piping or pipe supports may be damaged by the weight of the water.
- Ensure that flanges are opened and blinds are installed in a sequence that precludes air from being drafted into the flare stack. Any air ingress can create a flammable mixture, resulting in an explosion in the flare system.
- It is generally recommended that slight positive nitrogen pressure be maintained on the flare system during flare blinding operations and that workers are protected using a fresh air breathing system (hose line supplied or self-contained breathing apparatus [SCBA]).
- The sequence of blind installation and removal should be such that other process equipment is fully protected against potential overpressure until it is otherwise protected, for example, by venting to the atmosphere.
- Blinding a "live" flare header is not recommended and should only be done to prevent an even higher risk.
  - Even then, all production units' feed, or load, should be reduced and stable to reduce the risk of an unexpected release to the flare system.
  - This means that a detailed risk assessment should be undertaken, including developing additional precautions or mitigation measures and requiring a higher level of approval authority.

Considerations for Blinding Highly Toxic Systems or Units (Operations Team)

Certain process units or processes routinely utilize highly toxic chemicals on an ongoing basis. Additionally, operators are trained and have the appropriate PPE to handle normal upsets or small loss of containment events. These chemicals are safe when fully contained in the process; however, when opening flanges in preparation for equipment cleaning, inspection, or maintenance, these systems represent special and unique hazards.

When preparing to open or blind equipment that contains or has contained these materials, the system must be completely drained, purged, and neutralized to ensure all the materials have been successfully removed.

The most common of these chemicals are listed below:

- Hydrogen sulfide (H_2S)
- Hydrofluoric acid (HF)
- Chlorine (Cl)
- Ammonia (NH_3)
- Benzene (C_6H_6)
- Methyl Isocyanate (CH_3NCO)

Note: Please refer to the link in the Additional References section of this Chapter for the US OSHA List of Highly Hazardous Chemicals, Toxics and Reactives (Mandatory).

Verify System Is Deenergized and Ready for Blinding (Operations Team)

Operators should verify that the equipment and/or piping is completely isolated, vented of residual energy (to a zero-energy state), and fully depressured.

- Zero energy can be confirmed by ensuring that vents and drains on the equipment or piping circuit are fully open, unobstructed, and leak-free.
  Note: If a zero-energy state cannot be achieved, a risk assessment should be completed, and mitigation should be implemented to ensure a safe state before proceeding. A higher level of approval is required with mitigation.

The method of verification should include the following:

- To identify blind points, all blinding locations should be pre-tagged with a distinctive tag developed specifically for this purpose.
- The blind points should be discussed during the joint job site visit between the work permit issuer and permit acceptor.
  - The permit issuer is responsible for pointing out the tagged blind points to the permit acceptor.
    Note: It is critically important that the permit issuer ensure that the permit acceptor understands that neither he nor any member of his work crew has permission to open any flange not designated (tagged) for opening.
  - The permit acceptor is responsible for discussing these with the work crew at the job site.

- Verify or open bleeders and vents adjacent to or near the location where the blind will be installed.
  - Bleeders should be opened and unobstructed very near the blinding location.
  - For larger systems, multiple bleeders should be verified.
- Ensure that bleeders and valves remaining open during maintenance are properly tagged and shown on the energy isolation list.
- Verify energy isolation by attempting to start any electrical equipment such as mixers, pumps, compressors, fans, etc.
  - Attempted starts should be done from each starting point, including remote start switches.
  - If maintenance is to be executed on any electrical components or wiring, the electrical breaker(s) must be de-energized in compliance with lockout/tagout (LOTO) requirements.

Verify Integrity of the Primary Isolation Points (Operations)

The operator should verify that the valves at the primary isolation point isolations are holding without leakage through the valve seat:

- If the isolation valves are confirmed not to be leaking internally, gas testing of the vessel or equipment can begin.
  Notes: Several reminders about gas testing to confirm that work can begin without the use of breathing air or continued ventilation:
  - The initial gas test should verify sufficient oxygen, as most gas test machines need oxygen to be present before providing an accurate reading of flammables.
  - The oxygen test should verify at least 19.5% oxygen before proceeding, in accordance with OSHA standards.
  - This test is followed by testing for flammables (lower explosive limit [LEL]) and confirming a maximum of 10% LEL, and for hydrogen sulfide (H_2S), to verify a maximum of 20 parts per million (ppm) (as per OSHA standards).
- If these targets are outside the range, continued isolation and purging must be completed, followed by retesting.
  - If the valves cannot be confirmed to be holding or are found to be leaking, a risk assessment must be conducted. The risk assessment should identify additional mitigation measures or precautions that need to be implemented before work starts. This also requires approval from higher authorities and confirmation that the mitigation has been completed.

- Additional mitigation measures may include specialized PPE, extra fire protection, or using nitrogen or steam purge to prevent air from entering the equipment during blinding operations. If nitrogen purge is used during blinding, workers will be required to wear breathing air protection.

Issue Permit to Install Blinds (Permit Issuer – Operations/ Permit Acceptor – Craftsmen)

- Before starting any work, it is very important that a thorough joint site visit is conducted between the permit issuer and the permit accepter.
  - Visit and discuss the location of each planned pipe or equipment opening, including marking or tagging each flange to be opened with a special tag developed specifically for this purpose.
  - Operators should point out to the craftsmen each flange to be opened, discuss how the equipment is isolated, and verify that it is at zero energy.
  - This is an opportunity to discuss and resolve job site precautions, such as the use of inert gas, use of PPE, fire protection, ensuring any adjacent hot work is shut down, site evacuation, emergency or rescue procedures, and any other outstanding questions, issues, or concerns before any work is done in the field.

Nitrogen and Steam Precautions

- Ensure precautions to prevent asphyxiation if nitrogen or another inert gas is used. This may include requiring a hose line supplied breathing air or SCBA.
- Workers should be trained never to connect to a pipeline supplied with air for breathing purposes. Workers have died doing this because plant-supplied air is often backed up by nitrogen, or pipelines may be mislabeled and actually contain nitrogen.
- If steam is used for purging or to create a seal against the process, ensure that workers are adequately protected against exposure to the hot steam vapor or steam condensate.
  - Also, condensing steam can create a strong vacuum and damage process equipment or draw air into it. Ensure vessels with the potential for condensing steam are properly vented to prevent vacuum damage.

Final Decontamination (Operations)

The final decontamination can start once the isolation blinds are installed and the equipment is completely sealed off into separate vessels, piping circuits, or other individual sections. The choice of decontamination method or technique depends on the type and quantity of material remaining in the equipment.

Some questions that should be addressed include the following:

- What material needs to be decontaminated?
  - Is it hydrocarbon?
  - Is it liquid, or is it a sludge?
  - Is it an iron scale or another identifiable corrosion product? If it's iron scale, it may be pyrophoric and should be kept wet with water until removed from the vessel.
  - Is it polymer?
  - Is it a catalyst? Some catalysts may also be pyrophoric.

The path forward may be to execute the final decontamination using a vacuum truck or continue to purge and/or steam the equipment. Another option is to work with a reputable chemical vendor and circulate a chemical cleaning solution in the equipment or to use a chemical neutralizer to clear the vessel(s) of hydrocarbons and other contaminants to the state required to facilitate vessel entry, internal inspections, and maintenance. For example, a reputable chemical vendor can decontaminate pyrophoric materials from process vessels. I have witnessed an extreme case of using small explosive charges to remove hardened catalysts from process vessels. This extreme case requires multiple layers of safety procedures, trained personnel, and comprehensive safety precautions.

Gas Testing Requirements

Before opening the process equipment, a gas test is performed, testing for oxygen deficiency (O_2), flammables (LEL), and toxins (generally hydrogen sulfide (H_2S). This gas test aims to help determine the PPE requirements for those opening the process equipment. Tests for other toxic vapors may be required depending on the prior process service or other unique circumstances. For example, vessels or equipment in hydrofluoric acid (HF) service will require a test for HF or a vessel that may contain carbonyls will require a special test based on prior service.

The gas test is conducted before the equipment is opened, and the result establishes the type of PPE required to protect the workers. For example, will a SCBA be required for breathing air protection?

A gas test should be completed within two hours or less before work begins, possibly sooner and with more frequent testing depending on the job and process conditions. Continuous gas monitoring should be considered for higher-hazard tasks. Four-function continuous testing (O_2, LEL, H_2S, CO, and/or other toxics specific to the process) is recommended. If continuous gas testing is not used, it may be necessary to retest the work area later in the day during daytime heating, as higher temperatures can vaporize flammable liquids and have resulted in catastrophic explosions.

Most gas test machines require a test to confirm the presence of oxygen before performing the test for LEL (flammables). Many gas test machines, especially older ones, will not provide accurate LEL readings without sufficient oxygen (typically inaccurate in concentrations below about 19.5% oxygen). The LEL test measures flammable vapor, which is expressed as a percentage of the LEL. The H_2S instrument reads in ppm. The gas tester must be fully trained on the specific instrument being used. Additionally, the gas test device should be calibrated and bump-tested before testing. One more point: do not confuse LEL with total hydrocarbons. LEL is expressed as a percentage of the LEL, while total hydrocarbons are reported as ppm.

Another crucial consideration is that combustible liquids might be present even when a zero LEL gas test is conducted. If combustible liquids are present, the gas test machine may still read zero, even though hydrocarbons are present. This occurs because the combustible liquids have a flash point above 100 °F; at this temperature, the liquid doesn't vaporize. Therefore, an LEL reading does NOT indicate the presence of combustibles such as diesel at ambient temperature when no flammable vapor is present. However, if the combustible liquid is heated above its flash point, it will vaporize. Vaporization can lead to an accumulation of flammable vapor in equipment or a vessel, which has caused explosions and flash fires.

More on this point: diesel and other similar combustible products have a flash point above 100 °F (38 °C) and are inherently safer to handle at ambient temperatures. For example, the diesel flash point is typically greater than 125 °F (52 °C), whereas gasoline and light naphtha are highly volatile with flash points well below 0 °F (below −18 °C). Gasoline and naphtha will generate flammable vapors at ambient temperature. For example, the flash point for gasoline is about −45 °F (−43 °C) or lower. However, diesel and similar combustibles become just as hazardous as gasoline if heated to a temperature above their flash point!

Before authorizing hot work, always verify that the equipment is fully isolated, open to the atmosphere, and confirmed to be free of flammable materials, toxic vapors, and combustible liquids.

Please refer to Chapter 6, "The Key Elements of Effective Gas Testing," for additional details.

Other Worker Precautions

Due to the potential for "trapped" liquids, vapors, or pressure, no hot work should be conducted where flanges or equipment are being opened

To the extent possible, craftsmen should stand out of the line of fire and open the flanges away from them. They should never remove all bolts but slowly crack the flange open, checking for pressure, liquid, or vapor. They should be prepared to quickly reclose the flanges if pressure or liquid is present.

The area where the OPE work is being conducted should be restricted to only those involved in the OPE process. The work area should be barricaded to minimize access by other personnel and to help keep others away from the work area. Scaffolding should be built to "green tag" standards with toe boards and alternate egress built into the scaffolds and elevated platforms.

Because of the risk of static electricity, always ensure an electrically conductive drain pan or collection trough is available when opening flanges or accessing the vessel. These should be properly bonded and grounded, and used to collect any spilled residual product.

Any change in work execution or work permit requirements should drive a thorough review and job site reapproval before restarting work.

Be Aware of Parallel Process Trains

Parallel process trains, such as parallel heat exchanger trains or parallel desalting trains, represent unique hazards when clearing equipment for opening. For example, it is difficult to flush or wash through with flushing oil or chemical cleaning when equipment is laid out in a parallel train. Attempts to flush or chemical clean with the flow through two or more parallel trains have trapped the process fluid or the chemical cleaning solution in one or more parallel trains. This happens most often due to internal fouling of the equipment or by the configuration of the parallel train. This can create internal pressure and flow restrictions and limit or possibly totally restrict the flow through one or more parallel trains.

This has had catastrophic results; for example, at a Gulf Coast refinery where the product was trapped in a parallel heat exchanger train, the result was a loss of containment when one of the exchangers was opened. A devasting fire resulted in the killing of two of the craftsmen who were involved in the OPE task and severely injuring several others.

Always be cautious and suspicious when opening a parallel train after flushing or chemical cleaning, as they may be fouled internally and could have the product trapped. Always verify zero energy at all vents, drains, or open flanges, and never use hot work or permit any ignition sources nearby when opening these parallel trains. For example, never authorize hot work to burn bolts unless using the "hot bolting" technique (more on this below).

Fire Watch

The fire watch is a standby person assigned to be fully prepared with a charged fire nose and fire nozzle positioned and ready to respond quickly to "knock down" welding or cutting sparks and prevent the spread of fire. The fire watch should be positioned away from the line of fire of the equipment being opened or hot work but prepared for immediate response when needed.

What is generally not well known is that OSHA regulations require the fire watch to stay on station for 30 minutes after the hot work is finished (see OSHA 1910.252 (a) (III) (4) (8)). This is to address the possibility of a smoldering fire that could unexpectedly and quickly escalate into a full-scale event.

Hot Bolting

Although, as mentioned above, using hot work to open flanges is never recommended; there is one possible exception. If the flange bolts are seized, and due to the bolt condition, they cannot be broken with a hydraulic torque wrench, after careful consideration, this may require the work to be done partially with hot work. In this case, it may be necessary to use a process called "hot bolting." A word of caution: hot bolting means different things to different people. I am describing the correct procedure for hot bolting, which should be executed with caution.

Hot bolting replaces and tightens one bolt at a time without breaking the flange or seal. However, before attempting this highly hazardous task, a detailed hazard analysis will be required to identify and mitigate the serious potential for something to go wrong. This should also require a high level of review and approval before execution.

Before initiating hot work to remove a bolt, always verify that there are no combustible or flammable materials in or near the area, all the sewer covers are completely sealed, and fire protection and spark containment are in place. Always confirm with a gas test that no flammables are nearby. Hot bolting should also require an independent observer to ensure that only one bolt at a time is removed, that it is promptly replaced with a fresh bolt, that the new bolt is fully torqued before moving to the next bolt for removal, and that the seal is never broken. While not recommended, this has been done successfully when all precautions are in place.

Confined Space Entry

Vessel entry is always considered a high-hazard task covered by the OSHA regulation (29 CFR 1910.146), "permit-required confined spaces." This work requires special precautions, including a satisfactory gas test before entry, forced "clean air" ventilation for the duration that workers are in the confined space, a designated attendant to monitor the activities of the entrant(s), a personnel rescue plan, warning signs to keep others away, and if entrance covers are removed, the opening shall be promptly guarded by a railing, temporary cover, or other temporary barrier that will prevent an accidental fall through the opening and that will protect each employee working in the space from foreign objects entering the space.

The gas test should always include at a minimum the following: vessel oxygen content (percent O_2), flammable atmosphere (percent LEL), hydrogen sulfide (ppm H_2S), and any other chemicals specific to the process involved.

The regulation requires employers to identify all confined spaces in their workplace and determine if any are permit-required confined spaces. A permit-required confined space contains a hazardous atmosphere, an engulfment hazard, an entrapment or asphyxiation hazard, or other serious safety and health hazards. The standard also requires posting warning signs to keep others away from the activity.

A special work permit is required to authorize entry into a designated confined space. Each entrant must be tracked, and vessel entry and exit times must be recorded for all entries.

No person should break the plane of the process vessel access points with any part of their body until the gas test and all other provisions of the OSHA confined space entry standard are in place. This means no one shall put any part of their body into the confined space until an authorized permit issuer issues the entry permit.

Note: This is a summary. Please see the link to the OSHA regulation in the Additional References section of this chapter for additional details. Confined space entry is also covered in significant detail in the Wiley textbook by this same author, *Hazards of Nitrogen and Other Inert Gases: How They Can Be Safely Managed*.

Returning Equipment to Service

This book's primary focus is the hazards of opening process equipment (OPE) and how those hazards can be identified and mitigated. However, hazards also exist when returning process equipment to service, for example, restarting a process unit after completing a major turnaround or returning a process vessel, a pump, or a compressor back to onstream service after maintenance or repairs.

The Wiley textbook *Process Operations Safety: The What, Why, and How Behind Safe Petrochemical Plant Operations* covers these hazards in significant detail. In this book, we talk in detail about the hazards and how each hazard can be managed or mitigated. For example, unit leak testing to ensure system integrity, details on water freeing and air freeing all equipment before bringing in hydrocarbons, the availability of safety-critical devices, process stream disposition during unit startup, and a host of similar topics are covered as they relate to unit startup or the startup of individual pieces of equipment.

Quick Summary

Opening Process Equipment (OPE) is a hazardous activity. As a result, many catastrophic losses of containment incidents have occurred in the oil and gas industry and other industries. Many workers who were carrying out these activities have been killed or severely maimed in the resulting fires, explosions, and toxic releases.

All OPE tasks warrant detailed procedures, a rigorous permit-to-work process, a thorough joint job site visit, and frequent return site visits to ensure good communications, that hazards are properly identified and addressed, mitigated, and that work conditions have not changed.

Our role is to insist on the availability and use of these detailed procedures and practices for each OPE task. And, always, if anything looks out of place or doesn't look correct, we must stop the job, intervene, and make it right before restarting.

You could save a life!

End-of-Chapter Quiz

1 What is generally required by a process facility to develop a procedure for safely conducting a unit shutdown and decontamination procedure (according to US OHSA)?
Please choose all that apply.
- **a** The process and instrument drawings.
- **b** The materials of construction.
- **c** The process safety information.
- **d** The properties of chemicals used in the process.
- **e** The information on the unit's safety systems.

- **f** The list of low-point drains and high-point vents.
- **g** Equipment isolation and blinding points.
- **h** The locations of steam tracing, etc.
- **i** All of the above.
- **j** None of the above.

2 The preparation and opening of refining or petrochemical plant piping and other associated equipment can be very ______________and require specialized ______________, well-trained ________________, and specialized ______________ __________________ ______________.

3 What is the primary difference between the preparation of a crude unit and a hydrotreater when preparing for isolation, opening, and entry?

4 There are several general safety guidelines that should be followed when steaming process equipment when preparing for opening. Please name as many as possible.

5 When are we most susceptible to finding pyrophoric material that can result in an internal fire when opening process equipment?
Please select the correct answer.

- **a** We are most susceptible to this when the process service is sour and contains a high concentration of hydrogen sulfide (H_2S).
- **b** We are most susceptible to this when the process vessel is corrosive and has deposits of corrosion products in the vessel's bottom.
- **c** Chlorine can result in pyrophoric, and therefore, vessels in chlorine service are very susceptible to pyrophoric materials.
- **d** Pyrophoric materials are not an issue at oil and gas facilities.

6 Please list the most important precautionary steps to take when opening a process vessel identified as possibly containing pyrophoric materials.

7 Why is it important when planning blinding locations to set vessel blinds at the first flange off the vessel on all lines?

8 Before opening flanges for setting isolation blinds, whose responsibility is it to verify that all process equipment is properly drained and depressured and to ensure that the isolation valves are holding?
Please select the correct answer(s).

- **a** Maintenance planners
- **b** Maintenance craftsmen

c Area supervisors
d Operations

9 Why is it important to quickly install isolation blinds in fuel gas lines after shutting down a furnace, heater, or boiler?

10 What makes flare systems unique when preparing for blinding or opening?

11 Why is it important to sequence the blinding of flare lines to prevent air entry into a flare line or flare system?

12 When blinding or opening piping or process systems that previously contained a hazardous chemical, operators should verify that the equipment and/or piping are completely ______________, vented of residual __________ (to a zero-_________ state), and fully _____________.

13 What hazard exists when purging process vessels and other equipment with nitrogen, and what special PPE may be required?

14 What are the typical gas tests required before opening a process pipeline or vessel, and why are these performed?

15 When executing hot work in a process vessel and continuous gas testing is not done, a gas test should be completed within ________ hours or less before work begins, possibly sooner and with more frequent testing, depending upon the job and process conditions.

16 When gas testing a process vessel that is at, or near, ambient temperature (below 100 °F or 38 °C), a gas test will not detect a ______________ ______________ that may be present in the vessel.

17 Why is washing or flushing through parallel trains such as parallel exchangers difficult?

18 Where should the fire watch be positioned when using a “fire watch” while hot work is in progress?

19 The OSHA regulation requires that the fire watch remains on station for ______ minutes after the hot work is completed.

20 Hot bolting replaces and tightens ________ bolt at a time without breaking the flange or seal.
Hot bolting should also require an ________ __________ to ensure that only one bolt at a time is removed, that it is promptly replaced with a fresh bolt, that the new bolt is fully torqued before moving to the next bolt for removal, and that the seal is never broken.

21 A special work permit is required to authorize entry into a designated confined space. Each entrant must be tracked, and vessel entry and exit times must be recorded for all entries.
True or False.

22 What are the US OSHA requirements for executing an entry into a permit-required confined space?
Please name as many as you can.

Additional References

AirGas, Safety Data Sheet – Hydrogen Sulfide, https://cglapps.chevron.com/sdspds/SDSDetailPage.aspx?docDataId=457984&docFormat=PDF.

Center for Chemical Process Safety (CCPS) / American Institute of Chemical Engineers (AICHE), Guidelines for Preparing Process Equipment for Maintenance and Return to Service, Wiley Publishing Company, https://www.wiley.com/en-us/Guidelines+for+Preparing+Process+Equipment+for+Maintenance+and+Return+to+Service-p-9781394237913.

M. Darryl Yoes, Hazards of Nitrogen and Other Inert Gases: How They Can Be Safely Managed, Wiley Publishing Company, https://www.wiley.com/en-us/Hazards+of+Nitrogen+and+Other+Inert+Gases%3A+How+They+Can+be+Safely+Managed-p-9781394330270.

M. Darryl Yoes, Nitrogen Asphyxiation Case Studies, and Asphyxiation by Other Inert Gases, Wiley, https://onlinelibrary.wiley.com/doi/abs/10.1002/9781394330294.ch11?msockid=33d3a0651a4764a43c56b64f1bc9657c.

M. Darryl Yoes, Process Operations Safety: The What, Why, and How Behind Safe Petrochemical Plant Operations, Wiley Publishing Company, https://www.wiley.com/en-us/Process+Operations+Safety%3A+The+What%2C+Why%2C+and+How+Behind+Safe+Petrochemical+Plant+Operations-p-9781394271948.

Durlon Sealing Solutions, Hot Bolting Safety Precautions, https://www.durlon.com/the-dangers-of-hot-bolting/#:~:text=Hot%20bolting%20is%20the%20practice,is%20under%20reduced%20operating%20pressure.

HSSE World (Health, Safety, Security and Environmental), Systematic Approach for Safe Mechanical Completion, Pre-commissioning, Commissioning, and Startup, https://hsseworld.com/systematic-approach-for-safe-mechanical-completion-pre-commissioning-commissioning-and-startup/.

Occupational Safety and Health Administration (OSHA), 29 CFR 1910.252 General Requirements, https://www.osha.gov/laws-regs/regulations/standardnumber/1910/1910.252.

OSHA Fact Sheet – Procedures for Atmospheric Testing in Confined Spaces, https://www.osha.gov/sites/default/files/publications/atmospheric_test_confined.pdf.

SafetyCulture, Scaffolding Safety & Scaffold Tagging, https://safetyculture.com/topics/scaffolding-safety/.

Schneider, 4 Most Important Mechanical Isolation Methods of Process Piping & Equipment, https://as-schneider.blog/2021/10/13/4-most-important-mechanical-isolation-methods-of-mechanical-equipment/.

Specialty Steel Industry of North America (SSINA), Chloride Stress Corrosion Cracking, https://www.ssina.com/education/corrosion/chloride-stress-corrosion-cracking/.

State of California Code# § 5189.1, California Process Safety Management for Petroleum Refineries, https://www.dir.ca.gov/title8/5189_1.html.

The National Board of Boiler and Pressure Vessel Inspectors, Preventing Steam/Condensate System Accidents, https://www.nationalboard.org/index.aspx?pageID=164&ID=208.

UK Health and Safety Executive (HSE), Breaking Containment, Hydrocarbon Lines, https://www.hse.gov.uk/pubns/guidance/oce12.pdf.

UK Health and Safety Executive (HSE), Guidance on Permit-to-work Systems, https://www.hse.gov.uk/pUbns/priced/hsg250.pdf.

US Chemical Safety and Hazard Investigation Board (CB), Delaware City Refining Company, https://www.csb.gov/delaware-city-refining-company/.

US Chemical Safety and Hazard Investigation Board (CB), Final DCRC Bulletin, https://www.csb.gov/assets/1/20/final_dcrc_bulletin1.pdf?15839.

US Chemical Safety and Hazard Investigation Board (CB), Removal of Hazardous Material from Piping Systems, https://www.csb.gov/removal-of-hazardous-material-from-piping-systems/; https://www.csb.gov/assets/1/20/pipeopening bulletin.pdf?13882.

US Chemical Safety and Hazard Investigation Board (CSB), Key Lessons for Preventing Incidents When Preparing Process Equipment for Maintenance, https://acrobat.adobe.com/id/urn:aaid:sc:VA6C2:6870ab1b-26cf-4e71-9e97-759e5a9e34c4?viewer%21megaVerb=group-discover.

U.S. Department of Energy; Washington, D.C, Doe Handbook – Primer on Spontaneous Heating and Pyrophoricity, https://www.standards.doe.gov/standards-documents/1000/1081-bhdbk-1994/@@images/file.

U.S. Department of Labor, Occupational Safety and Health Administration (OSHA), Directorate of Science, Technology & Medicine, Office of Science and Technology Assessment, Deaths Involving the Inadvertent Connection of Air-line Respirators to Inert Gas Supplies, https://www.osha.gov/sites/default/files/publications/shib042704.pdf.

US Occupational Safety and Health Administration (OSHA), 29 CFR 1910.146 - Permit-Required Confined Spaces, https://www.osha.gov/laws-regs/regulations/standardnumber/1910/1910.146.

US Occupational Safety and Health Administration (OSHA), List of Highly Hazardous Chemicals, Toxics and Reactives (Mandatory), https://www.osha.gov/laws-regs/regulations/standardnumber/1910/1910.119AppA.

Westlake Chemical Corporation; Lake Charles, Safe Process Piping, Equipment & Vessel Opening Procedure, https://www.westlake.com/sites/default/files/Safe%20Process%20Piping,%20Equipment%20and%20Vessel%20Opening%20Procedure.pdf.

3

The Potential Hazards Associated with OPE

One of the most significant hazards is the unexpected release of pressurized fluids or gases, including toxic gases. The release of a hazardous product while opening process equipment (OPE) can severely injure personnel and can also result in significant fire and property damage, especially if the released materials are toxic or flammable. An explosion can also occur when the product finds an ignition source, causing even more severe damage.

Failing to properly isolate the system or vessel can cause a sudden release of pressurized fluids when the flange or access plate is opened. The pressure must be carefully verified to ensure it is at or very near atmospheric pressure before issuing the work permit for workers to open the equipment. This can happen due to leaking isolation valves or a cross-connection to another system that was overlooked during isolation. These systems must be thoroughly checked to identify any cross-connections or piping tie-ins that could lead to a product release when the system is opened.

Another hazard related to OPE is the fouling or blockage of process vents or drains. This can give a false impression that the system is at zero pressure when it is actually fully pressurized. When preparing equipment for opening, it is crucial to verify that each process vent or drain is open, unobstructed, and clear. Sometimes, this requires using a bleeder unplugging device to ensure the bleeder is free of obstructions. Figure 3.1 shows a commercially available bleeder unplugging device. This device is fully enclosed to protect workers from contact with process fluids and injuries. It features a pressure gauge that signals when the plug has been removed, a vent that safely releases pressure or allows steam to be injected into the bleeder, and it comes in various pressure ratings and configurations, which should be selected specifically for the piping system where it is used.

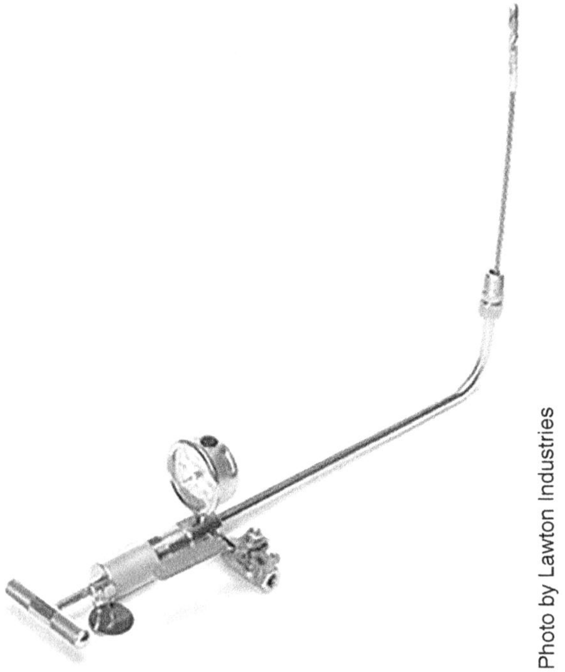

Figure 3.1 A commercially available bleeder cleaner.

A welding rod, as shown in Figure 3.2, should never be used for this purpose. I found this homemade unplugging device on an elevated structure at a European refinery near a control valve station. I recognized what it was and how it had been used because I had done this myself early in my career. Unfortunately, many workers have been injured while trying to unplug bleeders in this risky manner. The engineered bleeder unplugging device is the safe way to clear a blocked bleeder.

Sometimes, equipment elevation can cause trapped liquids to be released when opening a flange, allowing air to enter the system. There are cases where the piping system was drained, flushed, and steamed to remove residual material. Later, during the equipment opening, a significant spill of trapped hydrocarbons occurred due to a low point or a "belly" in the piping circuit that trapped the hydrocarbons.

I have also witnessed and am aware of other cases where a portion, or possibly the entire system, has previously been in a high-melting-point product that tends to harden when it cools, such as petroleum wax, a very heavy distillate, or a residuum or asphalt type of service, and others with a high melting point. In these cases, it is possible to form an internal product plug in the process system, which can melt when the line temperature rises above the product melting point, releasing trapped product.

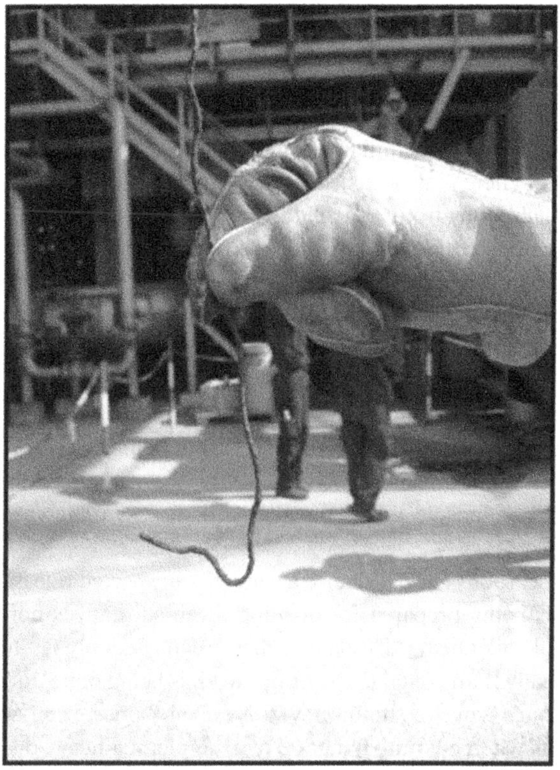

Figure 3.2 Photograph of a welding rod converted into an unplugging device and used to unplug a process bleeder. This can allow the released fluid or pressure to injure the worker and, therefore, should never be used for this purpose.

This is especially true in situations where the system relies on a single isolation valve that leaks beneath the valve seat. The hot product will seep through the valve very gradually, potentially heating and melting the internal product plug. This can happen due to a leaking isolation valve, daytime heating, or direct sunlight on the system. What makes this difficult to detect is that it can occur even when all vents and drains are fully open and seem unobstructed. When the system is heated above the product's melting point, the trapped heavy material melts suddenly, releasing the hydrocarbons trapped behind the internal plug. The hydrocarbon product, sometimes the hot product behind the isolation valve, is discharged, or a lighter product trapped behind the solidified heavy material is released. When this occurs, the release is abrupt and unpredictable, possibly causing a significant loss of containment and a major unit fire.

Another major hazard is the nature of the chemical that might be released. If it is flammable, such as light naphtha, gasoline, or liquefied petroleum gas (LPG), a sudden and unexpected release could create a vapor cloud and/or flash fire, potentially trapping workers and causing severe burn injuries. If it is toxic, like hydrogen sulfide, chlorine, ammonia, or similar substances, workers could suffer serious injuries or even death. There is also a risk of severe chemical burns or acid exposure from chemicals such as hydrofluoric acid, sulfuric acid, caustic solutions, phenol, or other toxic or corrosive chemicals. We will discuss this in more detail later, but verifying line or system pressure is critical, whether it is a pipeline or a process vessel; this must be done immediately before issuing the permit and confirming the system is at zero pressure.

I have already mentioned the system temperature. Often, steam is used to purge the system or vessel of leftover hydrocarbons. The temperature of the remaining condensate (hot water) must be confirmed as safe before opening. This must be done prior to issuing the work permit. Otherwise, workers could be severely burned by hot condensate when the flange or access plates are removed.

An accumulation of hydrocarbons can also cause a sudden, unexpected release and fire when equipment is being prepared for opening. Even substances normally considered combustible, like heavy distillate or diesel, can suddenly ignite into a dangerous fire, especially if any open flame or hot work is being done during the opening process. That's why we should never use open flames, such as cutting torches, to burn bolts when opening flanges. For example, if a heavy distillate is present and a gas test is performed before issuing a hot work permit, the test will most likely read zero flammables. This is because the hydrocarbons are at a temperature well below the product's flash point, so no vaporization occurs at these temperatures. However, when a torch heats the hydrocarbon, the temperature rises above the flash point, causing the material to vaporize quickly and create a flammable mixture that hot work can ignite. This exact scenario has resulted in worker fatalities.

Another hazard that has caused significant releases, major fires, and fatalities occurs when workers operating equipment inadvertently access the wrong piping circuit and open a flange that was not prepared for opening or entry. This often happens due to the operations team's inadequate line identification or a failure to conduct a proper start-of-job briefing in the field. This applies to all flanges and access plates, but is especially critical for work on long pipelines where pipes run directly next to other lengthy sections of similar pipelines in the pipe rack. Each flange should be marked with a unique "line opening" tag, typically designed in a distinctive color specifically for this purpose. This tag must stand out clearly from all other tags used in the field and should be clearly communicated to workers that it indicates the flange has been prepared for entry, and no flange without this tag should be opened.

Everything discussed so far leads to what I call a "bad day." However, as the situation develops, it only worsens. When a release causes an explosion and unit fire, we must always worry about the potential for escalation to an even larger incident. As more equipment is affected, further loss of containment of other process fluids and chemicals will inevitably happen. This can quickly spread to other process units and even to a large community event, with municipal orders for residents to shelter in place or evacuate. Besides the damage to property and personal injuries mentioned above, facilities must also consider the risk of escalation that could cause environmental damage or additional injuries on-site.

There is the potential for releases into the air, soil, and possibly into rivers, streams, lakes, etc. These may cause environmental damage, lead to environmental damage assessments, municipal fines or penalties, and harm the company's reputation.

Nitrogen Hazards

We use large quantities of nitrogen during OPE operations. This inert gas is often used to purge remaining hydrocarbons from piping and equipment before opening, and it is also used to remove air from equipment before reintroducing hydrocarbons during unit startup or when returning equipment to service. Nitrogen is inert. The Webster definition of inert states that it is lacking the power to move, very slow to move or act, sluggish, deficient in active properties, especially: lacking a usual or expected chemical or biological activity. The air we breathe consists of 78% nitrogen, 21% oxygen, and trace amounts of other gases. With this definition and the fact that the air we breathe contains 78% nitrogen, many people consider nitrogen to be completely safe to work with.

Nothing could be further from the truth. Nitrogen is deadly and has resulted in the deaths of many people, especially in our industry. This is especially true when nitrogen is vented into a confined space. In confined spaces or other areas with limited air circulation, nitrogen quickly displaces the air we breathe, and we can die from asphyxiation due to the lack of oxygen. One breath of pure nitrogen will shut the brain down; the victim stops breathing and will not restart unless they are resuscitated within about three to four minutes.

Some "rules of thumb regarding nitrogen":

- Always use certified Grade D breathing air in self-contained breathing apparatus (SCBA) equipment (Grade D is certified for breathing air and meets all the critical specifications). Manufactured or blended air should not be permitted at the facility.

- Each breathing air cylinder should be tested for O_2 content before use (the specification for oxygen content is between 19.5% and 23.5% by volume).
- Never use refinery air supply systems for breathing air; these systems are not certified as breathing air, may be contaminated, and often have pressurized nitrogen backup to protect the instrument air systems from low pressure. Deaths have occurred because the refinery system was labeled as air but was actually nitrogen (the refinery pipeline was mislabeled).
- Always confirm oxygen and H_2S levels with a valid gas test before entering a confined space.
- An authorized permit issuer must also issue a valid confined space entry permit before anyone enters a confined space.
- When purging equipment, ensure the vented nitrogen does not vent to a confined space or areas with limited air circulation.
- Additionally, barricades and warning signs should be placed in areas where nitrogen is vented when purging equipment, particularly near vessel entry points like manways.

Failure to follow these simple nitrogen safety rules has led to numerous fatalities in our industry. The deaths and injuries related to nitrogen have prompted me to write a book available on Amazon and other major retailers. The book is titled *Hazards of Nitrogen and Other Inert Gases: How They Can Be Safely Managed*. Therefore, I will not discuss these hazards further here, but if you are interested, I strongly recommend obtaining a copy of this book or accessing it online at wiley.com.

Be Aware of Normalization of Deviance

Diane Vaughan, author of The Challenger Launch Decision, introduced the concept of the Normalization of Deviance. She wrote about the US Space Shuttle Challenger disaster and the decision to launch. The Shuttle Challenger exploded on the cold morning of January 28, 1986, when the O-ring seals on the shuttle's solid rocket boosters failed, causing the external fuel tank to explode. This massive explosion destroyed the space shuttle and killed all 7 crew members about 73 seconds after launch from the Kennedy Space Center in Florida.

Normalization of deviance is a term used to describe a process where a person, a work crew, or an entire organization has developed a process that deviates from well-established procedures or norms, and the deviance has occurred so many times that it has become the accepted process or the new norm. The term normalization of deviance has been attributed to Diane Vaughan in her book describing the Space Shuttle Challenger disaster due to deviations from established procedures to the point where the deviations were accepted as the "new normal."

Chapter 10, Preventing "Normalization of Deviation" covers the loss of the space shuttle and the concept of normalization of deviance in more detail.

Special Awareness of Unique Hazards and Precautionary Measures That Should Be in Place

- Be continuously aware of the possibility of "normalization of deviation." This can slowly and unconsciously be implemented and accepted without normal reviews and approvals.
- Always verify the training and qualifications of craftsmen (employees or contractors) assigned to carry out the planned work. Ensure their leadership knows the job site hazards and how to summon for help or respond to an emergency.
- Ensure that an on-site joint job site visit is completed between the work permit issuer and acceptor during the issuance of the permit-to-work, with a discussion of all equipment opening locations. The operations group should identify, label, and prep all flanges or plates to be opened before this field visit.
- The locations and functions of the available safety showers and eyewash fountains should also be discussed during the joint job-site visit.
- Additional resources may be required to accomplish the preplanning, site safety inspection, joint job-site visit, permit issue and acceptance, and spark control. This should be considered in advance.
- When isolating equipment and where possible, always use double block valves with an open bleeder in between to detect internal leakage. If work is against a single isolation valve, recognize that potential leakage is possible and plans are in place to manage.
- Workers should be provided with a full suite of personal protective equipment (PPE) commensurate with the job site hazards and potential chemical exposure. In addition to standard worker-issue PPE, other PPE may include O_2 and toxic gas detectors, full face shields or mono-goggles, SCBA, chemical splash protection, fall protection, and a high-visibility vest.
- Ensure workers are aware of defensive positions and practices when opening flanges or access plates as follows:
  - Workers should always stand to the side of the flange out of the line of fire when opening the flange.
  - Always loosen bolts without removing any, and "crack" the flange open to check for liquid, vapor, or pressure before fully opening it. This allows craftsmen to reclose the flange if pressure or liquids are discovered.

- To prevent unexpected releases, piping or vessel openings that are expected to continue over several hours or days should be secured by installing slip blinds or blind flanges. Likewise, whenever equipment like pumps or compressors in flammable or toxic service are removed for repairs, ensure that blind flanges are installed or equipment is otherwise positively isolated.
- Always treat the opening of a flange for the purpose of installing or removing a slip blind or a figure 8 blind, or the removal of a blind flange as a line opening. They require a full work permit and full line isolation, venting, and draining.
- Be aware of potential chemical spills during tank preparation, especially during liquid removal or tank cleaning operations. Improper isolation or failure to fully drain a tank before opening can lead to a devastating release or chemical spill.
- Failure to fully isolate and depressure a high-pressure steam line before opening will lead to a large release of high-pressure steam and potential serious burns to personnel.
- When opening flanges or access plates on equipment or piping that has contained, or may still contain, flammable liquids or vapor, the worksite should be surveyed to identify and eliminate any potential ignition sources. This includes any hot work, vehicles, motorized equipment, electrical arcing devices, etc. Always check areas located above and below, especially downwind.
- When performing OPE on elevated surfaces or equipment located overhead, ensure that areas below the work area are properly barricaded and other workers are excluded from access by barricades. Ensure splash pans collect any residual material and ensure spark containment with fire water is in place as required.
- Good communication between the control room personnel, operations personnel, and field workers is critical before and during all phases of the work.
- It is very important to ensure that all personnel understand that in an emergency or in the event any hazardous product is detected, all work should be stopped immediately and restarted only after the site is reevaluated and all clear is given. The work permit should be reissued before work restarts.

End-of-Chapter Quiz

1 A significant hazard associated with OPE is the _________ or _________ of process vents or drains. This may give the false impression that the system is at _______ pressure when fully pressurized.

2 In a loss of containment event that results in a large unit fire, the situation may rapidly escalate due to additional equipment damage and even more product losses. This may escalate to involve other process ________, resulting in ____________ damage or personnel __________.

3 Another hazard associated with OPE is the fouling or plugging of process _______ or _________.

4 What are the advantages of using a bleeder cleaner to unplug a vent or drain instead of a welding rod?
Please select all that apply.
  - **a** The device is completely enclosed and protects workers from contact with the process fluid and a resulting injury.
  - **b** It is designed with a pressure gauge to indicate when the plug has been removed.
  - **c** It includes a vent that is used to vent pressure away safely or can be used to inject steam into the bleeder.
  - **d** It comes in various design pressures and configurations, which should be chosen specifically for the piping system where it is to be used.
  - **e** None of the above.
  - **f** All of the above.

5 Simply stated, what is the Normalization of Deviance?
Please select all that apply.
  - **a** A term used to describe an alternate process to the accepted process, which saves time when compared to the original process that the company specified.
  - **b** A term used to describe a process where the organization requires an alternative process that is different from the approved process but has been approved by the team leader.
  - **c** A term used to describe an alternative process that the team leader developed that saves both time and resources and is frequently used by all team members, although the organization does not require it.
  - **d** A term used to describe a process where a person or an entire organization has developed a process that deviates from the established procedure and has used it to the point where it becomes the accepted new normal.

6 When opening flanges or access plates on equipment or piping that has or may contain flammable products, the worksite should be surveyed to identify and eliminate any potential _________ ________.

7 It is very important to ensure that all personnel understand that in an emergency or in the event any hazardous product is detected, all work should be stopped immediately and restarted only after the site is reevaluated and all clear is given. The work permit should be reissued before work restarts.
True or False.

8 What should be used to collect residual material when opening process flanges or access plates?

9 Always treat the opening of a flange to install or remove a slip blind, a figure 8 blind, or the removal of a blind flange as a line opening. They require a full _______ ________ and full line _________, including equipment _________ and __________.

10 An accumulation of hydrocarbons may also result in a sudden and unexpected release and fire when equipment is being prepared for opening. Even what is normally considered a combustible, such as heavy distillate or diesel, can suddenly erupt in a dangerous fire, especially if any open flame or hot work is being done as part of the OPE.
True or False.

11 When process equipment is opened, there is the potential for release to the air, soil, and possibly to rivers, streams, lakes, etc. These may lead to ___________ damage and environmental _________ assessments, municipal _______ or penalties and damage to the company's ___________.

12 When a release results in an explosion and unit fire, what should we be concerned about when additional equipment is affected?

13 Which of the following is the better and safer alternative for equipment isolation?
Please select all that apply.

- **a** Isolation using double block valves with a bleeder valve in between to verify no leakage.
- **b** Isolation with a single block valve with downstream bleeder valve open to verify no leakage.
- **c** Isolation with double block valves with bleeder valves upstream and downstream of the valves open to verify isolation.
- **d** Isolation using a single isolation valve close to the equipment and another isolation valve a long distance also closed to provide secondary isolation.

14 One of the most significant hazards associated with OPE is the unexpected _________ of pressurized fluids or gases, including toxic gases.

15 Releasing a hazardous product while OPE can severely injure personnel and result in significant fire and property damage, especially if the released materials are ________ or __________.

16 Which two US Space Shuttles were lost due to normalization of deviance?

17 Sometimes, equipment elevation can result in trapped liquids released during flange opening, allowing air to enter the system.
True or False.

18 Why should we NOT use a refinery air system for breathing air?

19 When purging equipment, why is it important to erect warning signs and barricades where nitrogen is venting, especially near vessel entry points like manways?

Additional References

American Institute of Chemical Engineers (AICHE), Center for Chemical Process Safety (CCPS), Line Opening – Potential Hazardous Consequences, https://www.aiche.org/ccps/resources/tools/safe-work-practices/line-opening/potential-hazardous-consequences.

Center for Chemical Process Safety (CCPS), Guidelines for Preparing Process Equipment for Maintenance and Return to Service, https://www.wiley.com/en-in/uidelines+for+Preparing+Process+Equipment+for+Maintenance+and+Return+to+Service-p-9781394237890.

Diane Vaughan, The Challenger Launch Decision, https://www.google.com/search?q=diane+vaughan+challenger&sca_esv=1b74a963865752d4&ei=7OMgaIHDJpu_p84PnMq24AU&gs_ssp=eJzj4tLP1TcwKyhIK4o3YPSSSMlMzEtVKEssTc9IzFNIzkjMyUnNS08tAgDwRw0w&oq=Diane+Vaughan&gs_lp=Egxnd3Mtd2l6LXNlcnAiDURpYW5lIFZhdWdoYW4qAggBMgUQLhiABDIFEC4YgAQyBRAAGIAEMgUQABiABDIFEAAYgAQyBRAAGIAEMgsQLhiABBjHARivATIFEAAYgAQyCBAAGKIEGIkFMggQABiABBiiBDIUEC4YgAQYlwUY3AQY3gQY4ATYAQFIkW1QAFgAcAB4AZABAJgB2wKgAdsCqgEDMy0xuAEByAEA-AEC-AEBmAIDoALkcpgDALoGBggBEAEYFJIHCzMtMS42LTEuOS0xoAekDbIHCzMtMS42LTEuOS0xuAfkcg&sclient=gws-wiz-serp.

ExxonMobil – Mitchell Moloney – ExxonMobil Petroleum & Chemical B.V.B.A, Opening Process Equipment (OPE) – ExxonMobil Best Practice, https://refiningcommunity.com/presentation/opening-process-equipment-ope-exxonmobil-best-practice/.

Investigation of the Challenger Accident, Report of the Committee on Science and Technology, House of Representatives, Ninety-Ninth Congress, Second Session, https://www.govinfo.gov/content/pkg/GPO-CRPT-99hrpt1016/pdf/GPO-CRPT-99hrpt1016.pdf.

Lawton Industries, Inc., Lawton Angle Worm Bleeder Cleaners, https://www.lawtonindustries.com/.

Marathon Oil Company, Opening Process Piping & Equipment Field Reference Tool, https://lce.wp.marathonoil.com/wp-content/uploads/sites/4/2021/08/Opening-Process-Piping-and-Equipment-Field-Reference-Tool.pdf.

Mary R. Price and Teresa C. Williams, When Doing Wrong Feels So Right: Normalization of Deviance, National Library of Medicine / Center for Biotechnology Information, https://pubmed.ncbi.nlm.nih.gov/25742063/.

Report to the President, By the Presidential Commission on the Space Shuttle Challenger Accident, June 6th, 1986, https://sma.nasa.gov/SignificantIncidents/assets/rogers_commission_report.pdf.

TapRooT®, Normalization of Excellence – The Rickover Legacy – 18 Other Elements of Rickover's Approach to Process Safety, https://taproot.com/normalization-of-excellence-the-rickover-legacy-18-other-elements-of-rickovers-approach-to-process-safety/.

Texas A&M University / Texas A&M Stories, How the Columbia Shuttle Disaster Changed Space Travel, https://stories.tamu.edu/news/2023/02/01/how-the-columbia-shuttle-disaster-changed-space-travel/.

The INJAA Foundation, Construction Safety & Quality Consensus Guidelines, Basic Personal Protective Equipment (PPE), https://ingaa.org/wp-content/uploads/2018/10/18190.pdf.

The National Aeronautical and Space Administration (NASA),

Remembering the Columbia STS-107 Mission / Columbia Accident Investigation Documents, https://www.nasa.gov/remembering-columbia-sts-107/.

US Chemical Safety and Hazard Investigation Board (CSB), Final Report: Key Lessons for Preventing Incidents When Preparing Process Equipment for Maintenance, https://www.csb.gov/delaware-city-refining-company/.

US Occupational Safety and Health Administration (OSHA), Process Safety Management, https://www.osha.gov/sites/default/files/publications/osha3132.pdf.

4

When Are We Most Vulnerable to These Hazards?

First and foremost, we must recognize that fluids or gases can be released during all phases of the opening process equipment (OPE) process. In our business, these fluids may be highly flammable like petroleum naphtha or gasoline, extremely volatile liquefied petroleum gas (LPG), or highly toxic substances like hydrogen sulfide (H_2S), chlorine (Cl), hydrofluoric acid (HF), or ammonia (NH_3), among many others. Remember that what has happened before can happen again. This is crucial, and recognizing this is the first step in preventing future incidents at our facility. There are many lessons learned from past incidents, and we must learn from these and apply those lessons at our site. I believe it's much better to learn from incidents at other facilities and implement the lessons learned there, which can help prevent similar incidents at our own facility.

Please recall a few near misses at your facility where only luck prevented someone from being seriously hurt or a release happened without an ignition source being found. Of course, we should never depend on luck, and I often say that luck is not on our side. Everyone in the organization needs to recognize that without luck, we could have had a major incident, which means we are susceptible to a recurrence that could cause a "bad day."

The Pemex refinery in Deer Park, Texas, experienced what appears to have been an OPE incident in October 2024 when contract personnel were installing isolation blinds in a pipeline system. This tragic incident resulted in two contractors being killed and 13 others injured, along with a major release of hydrogen sulfide into the refinery and nearby communities. The Chemical Safety Board (CSB) is currently investigating the incident, and once the investigation concludes, we will have a clearer understanding of what happened. In addition to the two fatalities, the surrounding community of Deer Park was ordered to shelter in place, as was most of Pasadena, Texas. This caused significant disruption to the local community.

The CSB, in a November 20, 2024 update, reported, “During a maintenance activity at 4:23 pm, approximately 27,000 pounds of toxic hydrogen sulfide gas were released. In addition to the two fatalities and other serious injuries, the release resulted in a shelter-in-place order issued by officials in the nearby cities of Deer Park and Pasadena, Texas, lasting for several hours, and a portion of Texas State Highway 225 was closed temporarily.”

CSB Chairperson Steve Owens said, “Hydrogen sulfide is a highly toxic substance that can result in serious injury and death even at low concentrations. This dangerous incident resulted in the death of two workers and put other workers and the surrounding communities at very serious risk.”

In this same update, the CSB reported “that on the day of the incident, two contract workers at the refinery partially opened a flanged connection on piping containing hydrogen sulfide gas, which caused the release. One of these workers was fatally injured. The work was supposed to be done on a different, isolated piping segment about five feet away from the flange that was inadvertently opened. Two other contract workers from a separate company, who were working on equipment less than 250 feet away and downwind from the release, were also affected, with one fatally due to the hydrogen sulfide exposure.”

As stated earlier, we will have full details on how this incident happened, along with recommendations to prevent a recurrence when the CSB investigation is completed. We will also discuss this scenario in Chapter 7, “Controlling or Mitigation of OPE Hazards to Prevent Loss of Containment” and provide information on how this can be prevented from happening at your facility, at least based on what we currently know.

We are always vulnerable to a loss of containment that can result in a devastating explosion or fire that can injure people, result in extreme damage to plants and equipment, and damage the company’s reputation. An incident of this type can cost millions or even billions of dollars in equipment damage, environmental damage, claims, compensation, and penalties. This can happen suddenly and without warning, such as a brittle fracture failure. Brittle fracture is a catastrophic failure of the vessel that occurs when the vessel is pressurized when the metal temperature is below the minimum metal design temperature or MDMT. When pressurizing a process vessel, we should always verify that the metal temperature is above the MDMT on the vessel design plate or confirm this with the mechanical engineer. I discuss brittle fractures in more detail in the Wiley textbook *Process Operations Safety: The What, Why, and How Behind Safe Petrochemical Plant Operations*.

A release can start very slowly and then escalate rapidly, such as a leak from corrosion under insulation (CUI), although CUI failures have occurred with instantaneous failure and catastrophic results. We have management systems and programs, such as procedures and equipment inspections; however, well-trained operators are always our first defense against these failures. By following procedures closely, making thorough process rounds, and reporting anything that doesn't look right or needs attention.

Hazards of Transient Operations

From my perspective, backed up by the incident data, the most hazardous time when we are most vulnerable to an incident is during transient operations. For example, transient operations include shutting down or starting the facility or equipment, taking equipment out of service, preparing to open the equipment for cleaning or internal maintenance, conducting maintenance onstream, running plant tests, or being in an operating upset condition.

When opening equipment, it may have been in service for long periods, has contained hazardous hydrocarbons or chemicals, and may still contain these materials. During operation, no air was allowed to enter the equipment. The equipment must be drained, purged, steamed, or chemically cleaned to remove the hydrocarbons and chemicals before air can enter it for the first time, possibly in several years.

If air enters the equipment before preparation is complete, even a small amount of hydrocarbon or chemical residuals can easily form an explosive mixture inside the equipment being opened. Therefore, after draining the hydrocarbon or chemical, the equipment must be steamed, purged, or water-washed. If the equipment cannot support the weight of water, contains a catalyst, or for other reasons, it must be purged with nitrogen to create an inert environment before air is allowed in.

The reverse process occurs when we prepare equipment to return to process service. In this situation, the equipment contains air with about 21% oxygen content and atmospheric conditions. Essentially, all this air must be removed before hydrocarbons are introduced, as otherwise the hydrocarbons could mix with the air, potentially forming an explosive mixture.

BP Texas City, March 23, 2005: A Major Explosion During Unit Startup

The devastating explosion at BP Texas City on March 23, 2005 happened during a transient operation. The BP Isom unit was starting up after a scheduled outage. Although not directly connected to OPE, many of the CSB investigation findings relate to OPE; therefore, I included a brief discussion here to highlight those management systems and practices.

The BP Texas City explosion was a serious incident that resulted in 15 deaths and over 180 injuries, many of which caused life-changing harm, on March 23, 2005. The incident happened during a unit startup when a process unit fractionator was overfilled with hot, light naphtha. The naphtha was released from the atmosphere relief valves of the fractionator and rained down on the unit, catching many people completely off-guard. Many victims were working in wood-frame trailers near the fractionator when the explosion occurred. This incident caused billions of dollars in damages to plants and facilities, as well as victim compensation.

This incident appears to have resulted from normalized deviation, where the operators did not follow the written startup procedures when starting up the isomerization unit. The CSB identified numerous other causal factors that led to the fractionator column overfill and the release from the tower's atmospheric pressure relief valves. The US CSB found deficiencies in several operations and integrity management systems, such as safety leadership, operations integrity, operating procedures, process safety metrics, learning culture, management of change (MOC), and prestart-up safety reviews (PSSR).

This was demonstrated by deficiencies in BP's operations integrity management systems and key safety practices, some of which are listed below (not intended to be a complete list). I believe operations sites must ensure detailed procedures and practices are in place and strictly followed to maintain safety.

- Failure to ensure a detailed shift turnover process.
- Operating procedures were out of date and not followed by unit operators.
- The Isom unit overpressure release system vented to the atmosphere, not a closed system.
- Failure to ensure that critical instrumentation was operable before unit startup.
- Failure to clear noncritical personnel from the area before startup.
- Contractor maintenance trailers were allowed to be placed near the Isom Unit.
- There was a lack of competent supervision to oversee the unit startup.
- Console operators and others were fatigued due to working long hours and consecutive days.
- Workers were allowed to park an operating vehicle near the Isom unit while the unit was in startup mode (ignition source).

This incident clearly demonstrates why a robust and enforced process integrity management system is essential. Such a system must ensure procedures are current, all aspects of process safety are continuously monitored, and performance is tracked using key integrity indicators. The US Chemical Safety and Hazard Investigation Board conducted a thorough investigation and report on this incident, identified numerous discrepancies in BP's integrity management system, and presented findings indicating it was ineffective and that system requirements were not being enforced.

This is highlighted in the following quote directly from the CSB Final Report:

> 1.7.2.1 Safety Culture Recommendation On August 17, 2005, the CSB issued an urgent safety recommendation to the BP Group Executive Board of Directors that it convene an independent panel of experts to examine BP's corporate safety management systems, safety culture, and oversight of the North American refineries. BP accepted the recommendation and commissioned the BP U.S. Refineries Independent Safety Review Panel, chaired by former Secretary of State James Baker, III ("Baker Panel"). The scope of the Baker Panel's work did not include determining the root causes of the Texas City ISOM incident.
>
> The Report of the BP U.S. Refineries Independent Safety Review Panel was issued January 16, 2007. The Baker Panel Report found that "significant process safety issues exist at all five U.S. refineries, not just Texas City," and that BP had not instilled "a common unifying process safety culture among its U.S. refineries." The report found "instances of a lack of operating discipline, toleration of serious deviations from safe operating practices, and [that an] apparent complacency toward serious process safety risk existed at each refinery." The Panel concluded that "material deficiencies in process safety performance exist at BP's five U.S. refineries." 27 BP Texas City Final Investigation Report 3/20/2007.
>
> The Baker Panel Report stated that BP's corporate safety management system "does not effectively measure and monitor process safety performance" for its U.S. refineries. The report also found that BP's over-reliance on personal injury rates impaired its perception of process safety risks, and that BP's Board of Directors "has not ensured, as a best practice, that BP's management has implemented an integrated, comprehensive, and effective process safety management system for BP's five US refineries." The report's 10 recommendations to BP addressed providing effective process safety leadership, developing process safety knowledge and expertise, strengthening management accountability, developing leading and lagging process safety performance indicators, and monitoring by the Board of Directors the implementation of the Baker Panel's recommendations.

Many recommendations from the CSB BP Texas City Final Report and the Baker Panel have positively impacted process safety for other companies and workers at refineries and petrochemical plants. One of these outcomes was the creation of

API 752, a standard designed to protect personnel working in new and existing refinery and petrochemical plant facilities, including limiting the use of trailers in hazardous areas. Additionally, several excellent training videos were developed by the CSB and made available to the public.

Opening the Wrong Flange

Of course, other issues during the OPE process can lead to loss of containment of flammable or toxic materials, with tragic or catastrophic outcomes. Some of these include opening a flange still under pressure due to a plugged bleeder, opening the wrong flange or one that had not been prepared for entry, opening a process vessel containing pyrophoric iron sulfide scale resulting in a fire inside the fractionator or a fire within a storage tank. According to early reports from the US CSB, it appears that the incident at the Pemex refinery mentioned earlier occurred when the tradespeople accidentally opened a flange on the wrong pipeline, which was fully pressurized with a highly concentrated and deadly hydrogen sulfide (H_2S) gas.

An incident very similar to this one at the Pemex refinery happened at our facility long ago. At our site, workers were issued a work permit to install a 4-inch blind in a very long propane pipeline, which I estimated to be about 10,000 feet (3,048 meters) long. The line originated in the refinery and ended at an adjacent chemical plant. At the time, the two sites operated as separate facilities, but they shared a common emergency response team. I was a refinery superintendent and responded to the release during this incident. The chemical plant had a similar role for incidents at its site. However, whenever an incident occurred anywhere on the property, both emergency teams and their superintendents responded and took part in the response.

In this incident, the workers were issued the work permit in the control room, and since one of the contractors had installed this blind before, he thought he knew exactly where the blind was to go; therefore, there was no joint job site visit. When the workers arrived in the field, they removed all the bolts from the 4-inch flange and found the flange stuck to the pipe. They easily opened the flange with a hammer and a wedge, only to find it fully pressurized with liquid propane. There was an instantaneous and massive release of propane vapor, and of course, the workers were unable to reclose the flange. When liquid propane is released into the atmosphere from a liquid state, it rapidly vaporizes, auto-refrigerates, and expands to 270 times its original volume. Due to the auto-refrigeration, the spreading vapor cloud is heavier than air and quickly settles to the ground, forming a dense white fog. This is exactly what happened here.

Fortunately, there was no ignition, no explosion, and no fire. We were extremely fortunate; however, the potential for a catastrophic incident existed until the massive vapor cloud dissipated. Many lives could have been lost that morning, and I will never forget the scene and the potential for a massive explosion or major fire. Like most major incidents, even after a serious near miss, especially those with significant potential for personnel injury or equipment damage, generally, some great recommendations are generated to help prevent a recurrence. This one was no different. The recommendations from this incident are included in Chapter 11 and, along with many others, are designed to prevent this serious loss of containment incident. Several of those, specifically from this incident, are summarized as follows:

- The operations team is responsible for verifying that all flanges to be opened are verified to be at zero energy before the work permit is issued. All vents and bleeders should be open and proven to be unobstructed, and all cross connections must be fully isolated.
- Each line or vessel opening requires a joint job site visit between the permit issuer and those doing the work. If this is not feasible, a joint job site visit is held between the permit issuer and the craftsmen's representative. In this case, the representative is responsible for asking questions on behalf of the craftsmen and conveying all aspects of the job to the craftsmen.
- The flange(s) to be opened are identified with a tagging system using a uniquely designed tag specifically designed for this purpose and this purpose only. At our site, we implemented the requirement for a weatherproof and tamper-proof, psychedelically colored tag that is easily attached to the flange or equipment. Our operators quickly designated this tag as the "crazy tag." The tag is designed to be readily identified by the craftsmen to identify flanges and/or access plates to be opened. These tags should only be applied after the line has been fully prepared and demonstrated to be at zero energy to prevent flanges or access plates from being opened prematurely.

Internal Plugging Due to Solidified High-melting-point Products

Another thing to be aware of is that if you are preparing a system that has been in high-melting-point service, be cautious of any steam traps that may not be functioning properly. If the steam lines to and from the steam trap are found to be cold, heavier materials such as refined wax, a heavy refined distillate, or asphaltic or resid-type materials may solidify. This can prevent proper draining of the piping or equipment. As mentioned in Chapter 2, an internal plug may cause product and/or pressure to accumulate, which cannot be drained and will be released from opened flanges when the line temperature exceeds the product's melting point.

This occurred at a refinery in Canada, where workers were installing isolation blinds near a control valve station. When the flanges were opened, a slow flow of heavy residuum began to seep from the open flange. After a few minutes, the solidified residuum inside the piping melted, leading to a large release of extremely hot residuum and a major refinery fire. No one was injured in this incident, although significant equipment damage took place.

Preparation of Multiple Vessel Trains or Interconnected Vessels

Multiple trains are often found in heat exchanger systems where the process flow is split between several heat exchangers and then recombined when leaving them. This can also apply to some process vessels where, like exchangers, the flow is divided and runs in parallel paths through multiple vessels. Vessels arranged in parallel path trains have unique process safety risks, especially during preparation for opening. Commonly, the vessels are washed by flowing diluent hydrocarbons through them, followed by water and possibly high-pressure steam or nitrogen. Sometimes, chemical cleaning is performed by circulating a chemical solution through the exchangers. Despite all these efforts, hydrocarbon liquids may remain trapped in the exchangers and could be released into the atmosphere or ignite when opening the vessels.

This occurred at a Gulf Coast refinery, resulting in a tragic incident where two workers were killed, twelve others were injured, including five who were severely burned in a devastating fire. It's difficult to confirm that exchangers are clean, even after all the work mentioned above. Especially after a long period in service, the exchanger tubes may become heavily fouled, and even after chemical cleaning, it may be found that the cleaning was not effective because full flow was not achieved through the tubes.

In this incident, the reactor feed and effluent exchangers had completed a long run and were being cleaned and opened during the planned unit downtime. The exchangers were flushed with petroleum distillate and chemically cleaned by contractors, who circulated chemical solutions through the tubes and steamed them before attempting to open. The large exchanger studs were frozen in place, and despite using hydraulic torque wrenches, the bolts remained frozen. A decision was made to use hot work and burn off the bolts from the exchangers. Since all the inlet and outlet hydrocarbon lines were disconnected, no liquid was seen in the exchangers, and they had been gas tested, the team proceeded. However, when the bolts were burned, a flash fire occurred, resulting in two deaths and several others severely burned. Unfortunately, others were working on the exchangers or modifying the scaffolds at the time, and they were caught up in this tragic incident.

Hydrocarbon liquids trapped in the exchanger may originate from heavier diluent liquids used to flush the exchangers, typically a gas oil-type stream with a high flash point. Gas testing the exchangers will not indicate the presence of this liquid since it remains at or near ambient temperature. Due to its higher flash point, the heavier gas oil will not vaporize at ambient temperatures; therefore, no vapor will be present for the gas test. However, if hot work is performed to open the exchanger or if other ignition sources are nearby, this material could be easily ignited, leading to a major fire.

After an actual event or a near loss, the most important task is to ensure that the recommendations are fully documented and put into action. I believe there are people at sites where incidents have occurred, like those mentioned above, who follow the procedures and practices developed after the incident or even a serious near miss. However, they might not be aware of what happened in the past or how close we came to a disastrous event following a serious near miss.

Hot Bolting (Removing and Replacing One Bolt at a Time)

In an event like the Gulf Coast incident described earlier, where bolts are frozen or seized, there is a safer alternative that involves using hot work to remove the seized bolts. This is known as "hot bolting." However, this process should be considered only after exploring other options, and it must be performed carefully with precautions in place to properly manage the risk of igniting flammable or combustible materials.

First, I have learned that there are several different definitions of the term "hot bolting." I am referring to the process of using hot work or burning bolts with a torch, and the main difference is that we allow only one bolt to be burned at a time. The bolt is burned and then replaced with a new one, which is tightened or torqued to the manufacturer's specifications before moving on to the next stud for removal. It is important to note that no equipment will be opened during this process.

When performing hot bolting, the equipment must be completely depressurized and taken out of service for this process. The area should be screened to confirm that no flammable or combustible materials are nearby, and no other equipment is being opened in the vicinity. The area must be gas tested as part of the job site preparations, fire protection equipment should be prepositioned, and personnel who are not directly involved in the bolt removal should be kept away from the area. Process sewers must be covered and sealed to prevent vapors from entering, and similar precautions for typical hot work should be implemented before starting the hot work.

Once a hot work permit is issued by a qualified individual, we can proceed with burning one bolt at a time, reinstalling and tightening it before moving on to the next. After all bolts are replaced, the hot work is stopped, and the opening can be used with the normal method without hot work.

End-of-Chapter Quiz

1 We must recognize that we are vulnerable to the release of toxic or flammable fluids or gases during all phases of the __________ process.

2 Always remember that what has happened before can happen ________. This is extremely important, and we should recognize this is the first step in preventing it from happening at our facility.

3 The Pemex refinery in Deer Park, Texas, experienced what appears to have been an OPE incident in October 2024 when contract personnel were in the process of installing isolation blinds in a pipeline system. This was a tragic incident where 2 contractors were killed, 13 more were injured, and a major release of ________________ was released into the refinery and the surrounding communities.
Please select the correct answer(s).
a Hydrofluoric acid (HF)
b Benzene (C_6H_6)
c Hydrogen sulfide (H_2S)
d Chlorine (Cl_2)
e Monoethanolamine (MEA)

4 If the steam traps are discovered to be cold while draining fluids from low-point bleeders, liquids may be trapped in the system due to internal blockage caused by solidified products.
True or False.

5 What is unique about equipment like exchangers that are configured in a parallel flow path?
Please select the correct answer(s).

- **a** Following the process flow path through the exchangers is usually easy.
- **b** Parallel path equipment may contain trapped liquids that are difficult to clean and difficult to determine if they are present.
- **c** Trapped liquid hydrocarbons may be present even though a gas test reads 0% hydrocarbons.
- **d** Trapped hydrocarbons may be released when flanges are opened on exchangers configured in parallel paths.
- **e** None of the above.
- **f** All of the above.

6 Following an actual event or even a near loss, the most important thing is ensuring that the recommendations generated are ____________ fully and ____________.

7 When gas testing a process vessel or exchanger at ambient temperature or slightly above after opening, what product or material will NOT provide a flammable reading on the gas test machine?
Please select the correct answer(s).

- **a** Propylene at a temperature of 105 °F.
- **b** Diesel with a flash point of 145 °F.
- **c** Butane at a temperature of 65 °F.
- **d** Gasoline with a flash point of −45 °F.

8 Which of the following most correctly describes what is commonly referred to as "hot bolting"?
Please select the correct answer.

- **a** Depressuring and clearing the system, then burning all the bolts except four, one at each quadrant, then burning those one at a time.
- **b** Depressuring and clearing the system and heating the bolts with a torch before their removal.
- **c** Depressuring and clearing the system, burning one bolt at a time, reinstalling the bolt with a new one, and tightening it before moving to the next one.
- **d** Depressuring and clearing the system, performing a gas test to ensure that no hydrocarbons are present, and then burning all of the bolts.

9 If air is allowed to enter the equipment before preparation is completed, even if a very small amount of hydrocarbon or chemical remains, we can easily create an __________ ____________ inside the process equipment being opened.

10 What could occur within the piping when draining a high-viscosity piping circuit for hot work, especially if the steam traps are malfunctioning?

11 Which of the following is likely while attempting to clear a bank of heat exchangers which are in series flow:
Please select the most correct answers.
- **a** There is usually no problem clearing banks of exchangers that are in series flow.
- **b** Series flow exchangers can create pressure drop causing low flow when washing the tubes.
- **c** Sludge or coke material may accumulate in the exchanger tubes, causing significant amounts of liquid to accumulate in the exchangers.
- **d** Significant amounts of liquid may build in the exchangers because of the pressure drop caused by series flow exchangers.

12 When opening process equipment, what are some effective ways to prevent workers from opening the wrong flange or cutting into the wrong pipeline?
Please select the most correct answer(s).
- **a** Ensure a good walk-down of the job site with the workers, discussing exactly what equipment will be opened before the work permits are issued.
- **b** Ensure all flanges prepared for opening are properly tagged at the flange to be opened with a distinctive tag that is used only for this purpose.
- **c** For piping in long pipe bands, ensure that only the piping to be opened is painted with a distinctive colored stripe, and communicate to the workers that these are the only flanges to be opened.
- **d** Ensure that the job site is visited frequently and verify that the status of the ongoing work is in compliance with the work permit.
- **e** None of the above.
- **f** All of the above.

Additional References

AirGas, Safety Data Sheet – Hydrogen Sulfide, https://cglapps.chevron.com/sdspds/SDSDetailPage.aspx?docDataId=457984&docFormat=PDF.

API 752 Management of Hazards Associated with Location of Process Plant Permanent Buildings, https://www.api.org/oil-and-natural-gas/health-and-safety/refinery-and-plant-safety/process-safety/process-safety-standards/rp-752.

IChemE – Loss Prevention Bulletin 275, October 2020, Texas City Refinery Explosion – Safety Out of Focus, https://www.icheme.org/media/14611/lpb275_pg04.pdf.

US Chemical Safety and Hazard Investigation Board (CSB), BP America (Texas City) Refinery Explosion, https://www.csb.gov/investigations/completed-investigations/?F_InvestigationId=3515.

US Chemical Safety and Hazard Investigation Board (CSB), Incident Reports – Events Reported to the CSB Under the Accidental Release Reporting Rule Volume 1: Date: January 14, 2025, https://www.csb.gov/assets/1/6/csb_incident_reports_volume_one_2025-01-14.pdf.

US Chemical Safety and Hazard Investigation Board (CSB), Incident Reports – Events Reported to the CSB Under the Accidental Release Reporting Rule Volume 2: Date: March 12, 2025, https://www.csb.gov/assets/1/6/incident_reports_volume_2_2025-03-12.pdf.

US Chemical Safety and Hazard Investigation Board (CSB), Incident Reports – Events Reported to the CSB Under the Accidental Release Reporting Rule Volume 3: Date: July 22, 2025, https://www.csb.gov/assets/1/6/incident_reports_volume_3_2025-07-22.pdf.

US Chemical Safety and Hazard Investigation Board (CSB), PEMEX Deer Park Chemical Release, https://www.csb.gov/pemex-deer-park-chemical-release-/.

US Chemical Safety and Hazard Investigation Board (CSB), PEMEX Deer Park Chemical Release (Investigation Update), https://www.csb.gov/assets/1/20/pemex_investigation_1.pdf?17128.

US Chemical Safety and Hazard Investigation Board (CSB), Seven Key Lessons to Prevent Worker Deaths During Hot Work in and Around Tanks – February 2010, https://www.csb.gov/assets/1/17/csb_hot_work_safety_bulletin_embargoed_until_10_a_m__3_4_101.pdf?14329.

5

OPE and the Relevant US Occupational Safety and Health Administration Regulations

The US Federal Occupational Safety and Health Administration (OSHA) has incorporated regulations for conducting the "opening process equipment (OPE)"-type work into various existing federal standards rather than creating a specific or dedicated regulation for this worker activity. For example, to address requirements for equipment isolation and preparation for inspection, testing, or maintenance, OSHA has adopted the federal energy isolation standards into the "control of hazardous energy" standard, commonly known as the lockout/tagout standard, or 29 CFR 1910.147. Similarly, the agency developed the standard for entry into confined spaces in the regulation titled "permit-required confined spaces," or 29 CFR 1910.146. This approach has generally worked well, benefiting both industry and OSHA. However, with increasing incidents involving OPE or OPE incidents, there is a possibility that we may eventually see a specific standard outlining the requirements for this activity.

Meanwhile, I will do my best to outline the existing standards and other elements of what I refer to as "best practices" for safely performing this type of work. The regulatory standards will primarily be covered in this chapter. Additionally, I will include the best practices I've learned over the years, some of which I learned the hard way, in Chapter 7, titled "Controlling or Mitigation of OPE Hazards to Prevent Loss of Containment."

While I discuss US Federal OSHA regulations and requirements, which are implemented in many US states through the federal government, at least 29 OSHA-approved state plans operate statewide occupational safety and health programs. These state plans must have standards and enforcement programs that are at least as effective as federal requirements and may have different or more stringent standards. For example, California is well-known for its state regulations that appear to surpass the requirements set by the federal agency.

Certainly, I am mainly discussing the regulations in the US. Many foreign governments have also implemented regulations specific to process safety within their jurisdictions. While many of these align with OSHA regulations, I know that some overseas agencies have incident reporting and prevention requirements that go beyond US standards.

One of the OSHA standards that includes several detailed regulations specific to OPE-type work is the OSHA Process Safety Management Standard, or 29 CFR 1910.119, often called the "PSM standard." This is the main OSHA standard that requires employers to create, implement, and follow detailed procedures for handling or processing hazardous chemicals. The PSM standard is a mandatory document covering all industries, not just oil and gas, and aims to prevent the unexpected release of toxic, reactive, or flammable liquids and gases in processes involving highly hazardous chemicals. OSHA believes that regardless of the industry using these chemicals, there is a risk of accidental release if they are not properly managed. We have seen instances of this happening, and it's important to understand that this risk always exists with hydrocarbons and chemicals.

The PSM Standard is a comprehensive document and should be used as a reference for education and questions. The following is a high-level overview of the requirements that apply to OPE-type activities and are not meant to be exhaustive. The standard categorizes chemicals as toxic, reactive, flammable, or explosive, which could lead to a catastrophic release, and outlines certain precautions to prevent such incidents. These specific requirements are detailed in the OSHA standard.

The PSM Standard Includes the Following for Employee and Contractor Training, Operating Procedures, Including Startup and Shutdown Phases, to Manage the Associated Hazards

Employee Training: 1910.119(g)(1)(i) – Each employee presently involved in operating a process, and each employee before being involved in operating a newly assigned process, shall be trained in an overview of the process and in the operating procedures as specified in paragraph (f) of this section. The training shall include emphasis on the specific safety and health hazards, emergency operations including shutdown, and safe work practices applicable to the employee's job tasks.

Contractor Training – Employer Responsibilities: 1910.119(h)(2)(ii) – The employer shall inform contract employers of the known potential fire, explosion, or toxic release hazards related to the contractor's work and the process. 1910.119(h)(2)(iii) – The employer shall explain Yes contract employers the applicable provisions of the emergency action plan required by paragraph (n) of this section.

Contractor Training – Contract Employer Responsibilities: 1910.119(h)(3)(ii) – The contract employer shall assure that each contract employee is instructed in the known potential fire, explosion, or toxic release hazards related to his/her job and the process, and the applicable provisions of the emergency action plan. 1910.119(h)(3)(iii) – The contract employer shall document that each contract employee has received and understood the training required by this paragraph. The contract employer shall prepare a record containing the contract employee's identity, the date of training, and the means used to verify that the employee understood the training. 1910.119(h)(3)(iv) – The contract employer shall assure that each contract employee follows the safety rules of the facility including the safe work practices required by paragraph (f)(4) of this section. 1910.119(h)(3)(v) – The contract employer shall advise the employer of any unique hazards presented by the contract employer's work, or of any hazards found by the contract employer's work.

Standard Operating Procedures: The employer shall develop and implement written operating procedures that provide clear instructions for safely conducting activities involved in each covered process consistent with the process safety information and shall address at least the following elements: 1910.119(f)(1)(i) – Steps for each operating phase, including for Normal shutdown; and 1910.119(f)(1)(i)(G) – Startup following a turnaround, or after an emergency shutdown.

Other OSHA Standards That Include Specific Application to OPE Activities

Energy Isolation (Lockout/Tagout)

- 1910.147(a)(1)(i) – This standard covers the servicing and maintenance of machines and equipment in which the unexpected energization or start up of the machines or equipment, or release of stored energy could cause injury to employees. This standard establishes minimum performance requirements for the control of such hazardous energy.
- 1910.147(a)(3)(i) – This section requires employers to establish a program and utilize procedures for affixing appropriate lockout devices or tagout devices to energy isolating devices, and to otherwise disable machines or equipment to prevent unexpected energization,

start-up or release of stored energy in order to prevent injury to employees.

- 1910.147(c)(1) – Energy control program. The employer shall establish a program consisting of energy control procedures, employee training and periodic inspections to ensure that before any employee performs any servicing or maintenance on a machine or equipment where the unexpected energizing, start up or release of stored energy could occur and cause injury, the machine or equipment shall be isolated from the energy source, and rendered inoperative.
- 1910.147(c)(4)(i) – Procedures shall be developed, documented and utilized for the control of potentially hazardous energy when employees are engaged in the activities covered by this section. 1910.147(c)(4)(ii) – The procedures shall clearly and specifically outline the scope, purpose, authorization, rules, and techniques to be utilized for the control of hazardous energy, and the means to enforce compliance including, but not limited to, the following: 1910.147(c)(4)(ii)(A) – A specific statement of the intended use of the procedure; 1910.147(c)(4)(ii)(B) – Specific procedural steps for shutting down, isolating, blocking and securing machines or equipment to control hazardous energy; 1910.147(c)(4)(ii)(C) – Specific procedural steps for the placement, removal and transfer of lockout devices or tagout devices and the responsibility for them; and 1910.147(c)(4)(ii)(D) – Specific requirements for testing a machine or equipment to determine and verify the effectiveness of lockout devices, tagout devices, and other energy control measures.

Confined Space Entry

1910.146: This standard defines and establishes the requirements for a permit required confined space and a non-permit confined space as follows. Like the PSM standard, the following summarizes a very complex standard. Please refer to the complete standard for education or for questions.

- Permit-required confined space (permit space) means a confined space that has one or more of the following characteristics:
  1. Contains or has the potential to contain a hazardous atmosphere;
  2. Contains a material that has the potential for engulfing an entrant;
  3. Has an internal configuration such that an entrant could be trapped or asphyxiated by inwardly converging walls or by a floor which slopes downward and tapers to a smaller cross-section; or
  4. Contains any other recognized serious safety or health hazard.

- Non-permit-required confined space:
  A non-permit confined space is a space that does not contain or, with respect to atmospheric hazards, or the potential to contain any hazard capable of causing death or serious physical harm.
- The permit required standard also defines and describes the gas testing requirements and other potential personnel safety or health issues that may be present and must be addressed prior to entry.
- Also addressed are requirements and definitions for other safety precautions, including the posting of "DANGER – PERMIT-REQUIRED CONFINED SPACE, DO NOT ENTER" signs, the requirement and duties of an attendant or " hole watch," space ventilation requirements, pre-entry safety verification and entrant accountability, protection against falling objects or personnel into the space, a rescue plan with personnel readily available, the requirements for an entry work permit process prior to entry, and other support gear like two-way radios, lighting, ladders, etc.

Safe Work Practices and Specific Equipment Requirements

OSHA standards for safe work practices and for specific equipment requirements are embedded in many OHSA standards. For example, the following are examples of specific OSHA standards of this type:

Electrical safety	Hazard prevention and control
Emergency action plans	Injury and illness recordkeeping
Exit routes	Medical and first aid
Fire prevention	Personal protective equipment (PPE)
Fire safety	Respiratory protection
Hazard communication standard	Walking/working surface
Hazard identification and assessment	Worker participation

End-of-Chapter Quiz

1 The US Federal Occupational Safety and Health Administration (OSHA) has enacted and disbursed federal regulations for conducting process "opening process equipment"-type work into several existing federal regulations, rather than creating a special or dedicated regulation for this type of worker activity. For example, to cover the requirements for equipment isolation and preparation for inspection, testing, or maintenance, OHSA has enacted the federal requirements for energy isolation into the __________ ____________________________ standard affectionately referred to as the ________/________ standard, or 29 CFR 1910.147.

2 Many overseas and foreign governments have also enacted regulations that cover process safety specific to their ____________. While many essentially parallel the OSHA regulations, I am aware that some of the overseas agencies have requirements for incident reporting and prevention that ________ the US requirements.

3 Which OSHA Standard requires employers to develop, implement, and follow detailed work procedures and processes for handling or processing hazardous chemicals?

4 Which OSHA Standard defines and establishes the requirements for a permit-required confined space and a non-permit-required confined space?

5 Which of the following are covered or required by the OSHA PSM Standard? Please select all that apply.
   a The PSM Standard specifies the categories of chemicals that are defined as toxic, reactive, flammable, or explosive.
   b The PSM Standard defines the training required by each employee presently involved in operating a process and for each employee before being involved in operating a newly assigned process.
   c The PSM Standards require the employer to inform contract employers of the known potential fire, explosion, or toxic release hazards related to their work and the process and explain the applicable provisions of the emergency action plan.
   d The PSM Standard requires the contract employer to prepare a record that contains the contract employee's identity, the date of training, and the means used to verify that the employee understood the training.
   e The PSM Standard requires the employer to develop and implement written operating procedures that provide clear instructions for safely conducting activities involved in each covered process, consistent with the process safety information.
   f The PSM Standard does not require any of this information.
   g The PSM Standard requires all this information.

6 OSHA standards for safe work practices and for specific equipment requirements are embedded in many OHSA standards. Please list as many other OSHA equipment-specific standards as you can.

7 What sets apart an OSHA permit-required confined space from a non-permit-required confined space in the OSHA Confined Space Entry Standard?

8 The permit-required confined space standard details other requirements that must be in place prior to entry and for the full duration of an entry. Please list as many of these additional requirements as possible.

Additional References

British Health and Safety Health Executive (HSE), HSE Study Guide (Health, Safety, Environment), https://www.hsestudyguide.com/list-of-osha-standards/.

Safety Culture, OSHA Standards: A Guide to Health and Safety Compliance, https://safetyculture.com/topics/osha-standards/.

SC Training, 13 OSHA Training Requirements, https://training.safetyculture.com/blog/osha-training-requirements/.

US Occupational Safety and Health Administration (OSHA), Compliance Assistance Quick Start, https://www.osha.gov/complianceassistance/quickstarts/general-industry#:~:text=Hazard%20Communication%20Standard.,other%20requirements%20of%20the%20standard.

US Occupational Safety and Health Administration (OSHA), Control of Hazardous Energy (Lockout/Tagout), OSHA 29 CFR 1910.147. https://www.osha.gov/laws-regs/regulations/standardnumber/1910/1910.147.

US Occupational Safety and Health Administration (OSHA), Fact Sheet – PSM of Highly Hazardous Chemicals. https://www.osha.gov/chemical-hazards.

US Occupational Safety and Health Administration (OSHA), OSHA's A-Z Index, https://www.osha.gov/a-z.

US Occupational Safety and Health Administration (OSHA), Permit-Required Confined Spaces, OSHA 29 CFR 1910.146. https://www.osha.gov/laws-regs/regulations/standardnumber/1910/1910.146.

US Occupational Safety and Health Administration (OSHA), Process Safety Management of Highly Hazardous Chemicals, OSHA 29 CFR 1910.119. https://www.osha.gov/laws-regs/regulations/standardnumber/1910/1910.119.

US Occupational Safety and Health Administration (OSHA), Top 10 Most Frequently Cited Standards for Fiscal Year 2023 (Oct. 1, 2022, to Sept. 30, 2023), https://www.osha.gov/top10citedstandards.

6

The Key Elements of Effective Gas Testing

The importance of effective gas testing cannot be overstated. It is one of the most vital activities at a petroleum refinery or petrochemical plant. Gas testers must be trained on the specific equipment they will use at the facility and the types of tests that will be conducted. They must also pass a rigorous written exam and a field demonstration, then obtain certification from the facility before beginning gas testing. Additionally, they need periodic refresher training at specified intervals, usually annually, to ensure their qualifications remain current. It is best practice for the facility to keep a list of certified gas testers. This is essential to ensure all personnel conducting gas tests understand their role in preventing process safety incidents, which could include worker fatalities caused by asphyxiation, explosions, or fires. The key points outlined here provide minimal guidance that should be included in gas tester training.

To highlight the importance of gas testing, the US Coast Guard (USCG) issued a safety bulletin in September 2024 about a tragic welding incident that caused a fuel tank explosion on a commercial fishing vessel. The incident happened during maintenance, resulting in injuries and one death. The USCG bulletin stated that OSHA, under 29 CFR 1915 Subpart B, requires workers performing hot work near fuel tanks to make sure the area is certified as safe. OSHA calls for a "gas free" certificate, which is usually our hot work permit, to verify the space has no flammable vapors. The bulletin emphasized that "compliance with these measures is essential to avoid dangerous explosions or fires during welding or other hot work."

The USCG bulletin also reported that the National Fire Protection Association (NFPA): NFPA standard 51B outlines comprehensive safety guidelines for welding, cutting, and other hot work activities, emphasizing fire prevention and protection. This includes the requirement for hot work permits, fire watches, fire suppression equipment, and additional safety measures in confined spaces where the risk of explosion may be higher. Of course, in a petroleum refinery, we also take these precautions and procedures very seriously due to the potential for loss of containment that could lead to fires or explosions at our facilities. Preparing equipment for opening or hot work requires employee training, careful planning, attention to detail, and strict adherence to established practices and procedures.

Petroleum refining and petrochemical facilities contain many flammable, toxic gases, and asphyxiants. From my experience, aside from flammable gases, nitrogen and hydrogen sulfide (H_2S) are the two most deadly, having caused many worker fatalities. Nonetheless, most, if not all, of these tragic incidents were entirely preventable.

The Hazards of Nitrogen and Hydrogen Sulfide

Nitrogen is dangerous because it displaces the oxygen we need to breathe. It is completely odorless, and in most cases, the victim collapses, stops breathing, and dies within a few minutes. We use a lot of nitrogen in refineries and chemical plants. It is used to displace tank trucks during offloading, to gas blanket storage tanks to prevent flammable atmospheres, to conduct tightness tests on process equipment, and to purge equipment to make it gas-free or to remove air before returning it to service. These are just a few common uses of nitrogen; there are many other applications for this inert gas.

Nitrogen surrounds us. Air is composed of 78% nitrogen, 21% oxygen, and about 1% other gases. The danger isn't the nitrogen; the displacement of oxygen is what causes asphyxiation. When oxygen levels fall below 19.5%, we begin to see the effects of oxygen deficiency, and as levels decrease further, coordination and judgment become impaired. An even greater decline can lead to collapse and death.

H_2S is easily detectable at low concentrations. Most refinery workers recognize this and have smelled the strong rotten egg odor near process equipment operating in sour service. I call H_2S a "devious" gas because, at moderate levels, around 150–200 parts per million (ppm), it can deaden the olfactory nerves, causing the victim to believe the gas has disappeared or dissipated. As the concentration increases, the victim may no longer be aware that H_2S is present, and death can occur at levels as low as 500 ppm. One breath at levels of 1,000–2,000 ppm can kill instantly. OSHA's maximum permissible exposure limit (PEL) is 20 ppm, and the immediately dangerous to life and health (IDLH) level is just 100 ppm. To put this hazard in perspective, most petroleum refineries have some process equipment operating at concentrations ranging from 70% to 90%, or 700,000 to 900,000 ppm.

H_2S is not only highly toxic but also highly flammable. Its wide flammable range makes it easily ignitable by sparks and flames. H_2S has been known to ignite from static electricity in process fluids. Refinery workers have suffered serious injuries from H_2S fires, and many have been killed because of the gas's toxic properties.

While these are two of the most common gases involved in injuries or deaths of refinery workers, they are not the only one's workers may be exposed to or encounter during their work. Gas testers must be well-trained in these hazards and stay alert while testing piping, vessels, and other equipment. In addition to testing for flammable gases (LEL), gas testers should be trained in the following, if they are specific to their unit or area (this is not intended to be a complete list):

- Carbon monoxide (CO)
- Hydrogen fluoride (HF)
- Sulphur dioxide (SO_2)
- Nitrogen dioxide (NO_2)
- Chlorine (CL_2), and
- Ammonia (NH_3)

Flammable Limits for Typical Hydrocarbons

The flammable range, also called the flammable limit, explosive range, or explosive limit, specifies the mixture of fuel and air that can ignite or potentially explode. It is the ratio of fuel to air that can lead to ignition or explosion, known as the lower explosive limit (LEL) and the upper explosive limit (UEL). The LEL is the lowest concentration of gas in air that can ignite when exposed to an ignition source such as an open flame, arc, or spark. Below this level, an explosion or fire cannot occur. The UEL is the highest amount of hydrocarbon in the fuel/air mixture that will ignite when exposed to an ignition source; above this level, an explosion or fire cannot happen. The difference between the LEL and UEL is the flammable range. This range varies for each hydrocarbon and depends on the molecular makeup of the carbon and hydrogen molecules.

If the fuel/air mixture is below the LEL, it is too lean to ignite, or above the UEL, it is too rich to ignite, when exposed to an ignition source. However, when the mixture is between the LEL and UEL, it can ignite or possibly explode upon exposure to a spark, flame, or arc. Most hydrocarbons fall within a range of about 1–2% LEL to about 7–9% UEL. Some substances, however, have wider flammable ranges, such as hydrogen (4–75%), hydrogen sulfide (4–44%), acetylene (2–100%), and carbon monoxide (2–74%). When gas testing, we are measuring the LEL, which is the amount of flammable material that will ignite when exposed to a flame, spark, or arc.

Gas detectors measure a hydrocarbon's concentration as a percentage of its LEL. For example, the flammable range for methane is from 5% LEL to 15% UEL. Therefore, a space with no methane gas would read 0% LEL on the detector. Conversely, an environment with 5% methane would show 100% LEL. Essentially, when hydrocarbon vapors reach the LEL and an ignition source is present, it can lead to an explosion or at least a fire.

Gas Tester Training

Gas testers must be trained to perform all initial tests from outside the confined space. For example, a gas tester can use a tether or rope to lower the sample tube into the test area. Gases tend to stratify in process vessels and excavations, so it is advisable, especially when testing in large process vessels or deep excavations, to conduct tests at multiple levels or different vertical points. For instance, test the environment at the bottom, middle, and top of the space. Testing every 3–4 feet vertically is recommended.

Gas testers must be trained to survey the vessel and test all spaces that may contain hydrocarbons or sludge trapped inside, which can release vapors during hot weather or leak into areas where hot work is being performed. For example, naphtha and gasoline have been detected in the pontoons of storage tanks, causing internal fires due to pontoon leaks. These areas must be inspected and tested before issuing hot work permits to ensure safe working conditions.

Gas testers must have properly calibrated gas testing equipment and perform daily bump tests to ensure accurate results. Bump testing uses calibration gas to verify the device's accuracy. The equipment should include a sample tube or test hose long enough to reach remote areas where flammable gases might be trapped in a process vessel or excavation. Training should ensure that gas testers understand the importance of testing these remote areas and that they extend test times to allow the gas to reach the test device.

When gas testing a process vessel, gas testers must follow the correct sequence to perform the test. Typically, this starts with testing for oxygen (percent O_2), then hydrocarbons (percent LEL), and finally hydrogen sulfide (H_2S ppm). In petroleum refineries and petrochemical plants, these tests are conducted for all gas testing in process vessels, excavations, and other areas. Additional tests may be needed for chemicals specific to the process unit or area where testing occurs. For example, these may include hydrofluoric acid, chlorine, ammonia, or other chemicals unique to the process unit.

If gas test results show that the area is unsafe, workers should not be allowed to enter until the space is made gas-free and successfully gas tested. Gas testers must be trained to respond immediately if they find an area is unsafe. For example, if workers are already inside, they must be evacuated quickly and not allowed to reenter until the area or excavation is safe and confirmed with a repeat gas test.

As testing continues, results should be recorded directly on the work permit for the task being performed. In some cases, the initial gas test may be followed by continuous gas monitoring with a multi-gas detector. These multi-test devices are designed to constantly monitor oxygen levels and various toxic chemicals, providing both audible and visual alarms if test limits are exceeded. For example, a typical multi-test device might check for oxygen, hydrogen sulfide, hydrofluoric acid, chlorine, and other toxic chemicals every few seconds, sounding alarms if limits are exceeded. These devices significantly enhance the safety of personnel working in confined spaces or areas where hydrocarbons or chemicals could be present. I strongly support continuous gas testing.

The training should include guidance for gas testers to perform a visual check of their gas test equipment before each field use. This inspection helps identify signs of wear or damage to the equipment and the test tube. They should also be trained to inspect the vessel or process area when approaching the test site to ensure proper vessel preparation for entry or hot work. For example, they should verify that the vessel is prepared and properly isolated from other equipment, that manways or access points are open and unobstructed, and that there are no trapped hydrocarbons, sludge, or deposits that could release hydrocarbons. After this inspection, they can proceed to test the equipment in the correct sequence: oxygen, hydrocarbons, hydrogen sulfide, and any other chemicals specific to the process.

As a reminder, when using a longer test tube to access remote areas within a vessel, the gas tester should be trained to allow enough time for the gas test device to draw the sample into the testing equipment. They must keep the test tube in the testing area for several seconds to help ensure accurate test results.

For areas or tests without continuous monitoring, it is advisable for the gas tester to periodically return and recheck the equipment to ensure no flammable gas has accumulated. For instance, it is good practice to perform return gas testing immediately after workers take a break and before they resume work, after lunch, following any work stoppage, or if the confined space has been vacated for more than 30 minutes.

I have covered continuous gas monitoring in other chapters; however, because of its importance and benefits, I'll briefly mention it again here. Continuous gas monitoring is a crucial part of process safety because it can detect multiple gases at once in real time and activate both audible and visual alarms before gas levels become dangerous. The technology for continuous gas monitoring has significantly improved in recent years, and options are now available for many hazardous gases. Additionally, the availability and affordability of equipment have increased as the technology has progressed. As I mentioned before, I am a strong supporter of continuous gas monitoring and believe it can save lives.

All confined spaces must be tested according to these guidelines before anyone enters. This includes all confined spaces such as excavations or pits deeper than four feet, railcars, tank trucks, storage tanks, or other process vessels. Over the past year, while writing several process safety textbooks, I have read many cases where workers, believing they would be inside the vessel, tank truck, or railcar for only a few minutes, entered without a gas test and lost their lives inside the space. In about half of these cases, another worker rushed in to save the victim, only to also perish there. What is most tragic about this is that in many of these instances, the company involved had no procedures for confined space entry, no gas testing equipment available to workers, and no training for confined space entry. Gas testing, when done effectively, saves lives.

Permissible Limits for Vessel Entry or Hot Work

Always follow your facility's guidance for acceptable gas test results. However, if your facility does not have approved guidance, the following is suitable for most circumstances.

- Oxygen content: Oxygen level must be between 19.5% and 23.5%.
  Note: OSHA standards define oxygen levels below 19.5% as oxygen-deficient and immediately dangerous to life or health, while atmospheres above 23.5% are considered oxygen-enriched and present a significant fire hazard. OSHA Shipboard regulation is a maximum O_2 of 22% (1915.11).
- Flammability: LEL should be 10% or below to be considered safe. OSHA requires concentrations to stay below 10% of the lower flammable limit (LFL) or LEL.
  Note: I prefer the LEL to be 0% before issuing a hot work permit.
- Toxicity: Tests for toxic gases such as hydrogen sulfide (H_2S) and carbon monoxide (CO):
  - Hydrogen sulfide should be less than 20 ppm.
    Note: The OSHA standard for general industry is a ceiling limit of 20 ppm and a peak limit of 50 ppm for no more than 10 minutes during an 8-hour shift.
    The National Institute for Occupational Safety and Health (NIOSH) recommended PEL is 10 ppm.
    Note: I prefer a maximum concentration of 5 ppm or less.
  - Carbon Monoxide should be less than 35 ppm, and immediate evacuation at 50 ppm.
    Note: The OSHA 8-hour PEL is 50 ppm averaged over an 8-hour workday. The NIOSH recommended PEL is 35 ppm.

There is one more important point I want to emphasize about gas testing – one that has led to fatalities. When a vessel has been gas tested and shows 0% LEL, it doesn't always mean the danger is gone. At a US Gulf Coast refinery, workers were burning the bolts on a heat exchanger to open the equipment when a massive fire broke out. This incident resulted in two deaths and several others suffering severe, life-changing burn injuries. So, how did this happen, since the exchanger had been gas tested and declared free of flammable vapor?

In this tragic case, the exchanger still contained liquid hydrocarbons similar to diesel fuel but with a higher flash point. Fuels with a flash point above 100 °F (38 °C) are classified as combustible, while those with a lower flash point, like gasoline, are considered flammable. A combustible fuel with a higher flash point will not vaporize at ambient temperature, so without vapors, the gas test machine cannot detect the presence of hydrocarbons. However, when these liquids are heated by exposure to higher temperatures – whether from the sun or heat from work – they begin to vaporize and can ignite. In this tragic incident, the vapors ignited due to the hot work.

The main lesson here is that a successful gas test doesn't mean the vessel is free of hydrocarbons. It only shows that no vapors were present during testing. The gas tester needs to understand this and consistently check for liquid hydrocarbons. If any are found, they should notify the operations team and ensure a work permit isn't issued until the issue is fixed.

End-of-Chapter Quiz

1 Gas testers must be ____________trained on the specific equipment they will use at the facility. They must also pass testing and obtain ____________ from the facility before beginning gas testing in the field.

2 Before anyone enters confined spaces, such as excavations or pits deeper than ____________, they must be tested following the gas testing guidelines.
Please select the correct answer.

a 18 inches (1½ feet)
b 24 inches (2 feet)
c 48 inches (4 feet)
d 60 inches (5 feet)

3 All process vessels must be gas tested before they are entered, which includes the following.
Please select all that apply.

a Process fractionation column.
b Process chemical mixing drum.
c Excavation greater than 3 feet in depth.
d 45 K barrel storage tank.
e 6-foot tall 200-barrel open-top water tank.
f All of the above.

4 To ensure their qualifications remain current, certified gas testers require periodic refresher training at specified intervals, typically ______________.

5 Bump testing of gas testing devices involves using calibration gas to verify the testing device's ___________ and should be done ________.

6 Many people have died when entering railcars and tank trucks because the atmospheres were not __________ before entry. Others also died while attempting to _________ the first victim.

7 OSHA standards define oxygen levels below _______% as oxygen-deficient and immediately dangerous to life or health, while atmospheres above _________% are considered oxygen-enriched and present a significant fire hazard.

8 When the gas tester is sampling and using a longer sample tube to reach the depths in a process vessel, they should be trained to allow enough time for the gas test device to draw the sample into the test equipment. They must keep the test tube in the testing area for ____________ __________ to help ensure accurate test results.
Please select the correct answer.
- **a** About 20 seconds
- **b** No longer than one minute
- **c** Five minutes
- **d** Several seconds

9 For areas or tests without continuous monitoring, it is advisable to have the gas tester periodically return and __________ the equipment to ensure no _________ gas has accumulated.
For example, it is good practice to perform return gas testing immediately after workers take a _________ and before they resume _________ after lunch, after any work stoppage, or if the confined space has been vacated for more than about ________ minutes.

10 What are the benefits of using continuous gas testing?
Please select the correct answer(s).
- **a** Continuous gas monitoring tests continuously for the presence and concentration of multiple gases, typically performing tests every few seconds.

- **b** Continuous gas monitoring provides audible and visual alarms when gas is detected above preset values.
- **c** Continuous gas monitoring does not depend on human intervention or testing.
- **d** The technology and availability of continuous gas monitoring equipment have greatly improved over the past few years.
- **e** All of the above.

11 Please describe the sequence that gas testers should use when conducting gas tests on process piping vessels, excavations, etc.

12 When gas testing a floating roof storage tank for planned hot work, for example, welding inside the tank, why is it necessary to gas test the pontoons on the tank in addition to the interior of the tank?

13 If a gas tester performs a recheck while work is in progress and the gas test results indicate that the space is unsafe, what actions should the gas tester take?

Please select the correct answer.

- **a** The gas tester should immediately notify their supervisor that the space is unsafe.
- **b** The gas tester should caution the workers and inform them of the gas test results.
- **c** The gas tester should call for a recheck of the space with a different gas test device.
- **d** The gas tester should return to the operator shelter and conduct a bump test of his gas test device.
- **e** The gas tester should immediately stop the work and evacuate the space, moving the workers to a safe area.

14 Propane has a flammable range of between 2.2 and 9.6%. If the gas tester is sampling a propane vessel and there is a reading of 100% LEL, what is the minimum concentration of propane present?

15 At what percentage LEL is there likely to be an explosion or fire if a source of ignition, such as an open flame, an arc, or a spark, is nearby?

Additional References

CO2Meter Gas Measurement Specialists, Carbon Monoxide Levels and Exposure Limits Chart, https://www.co2meter.com/blogs/news/carbon-monoxide-levels-chart?srsltid=AfmBOorVEuZtvU1KUcxylp8h0D3GnhSPF44xXrjK3CMdXMGPjG9FaCtX#chart.

M. Darryl Yoes, Hazards of Nitrogen and Other Inert Gases: How They Can Be Safely Managed, Wiley Publishing, https://www.amazon.com/Hazards-Nitrogen-Other-Inert-Gases/dp/139433026X.

M. Darryl Yoes, Hydrogen Sulfide (H_2S) Hazards: Risk Recognition and Prevention, Wiley Publishing, https://books.google.com/books/about/Hydrogen_Sulfide_H2S_Hazards.html?id=xltC0QEACAAJ.

US Occupational Safety and Health Administration OSHA Confined Space Entry Standard, https://www.osha.gov/laws-regs/regulations/standardnumber/1910/1910.146.

US Occupational Safety and Health Administration OSHA Fact Sheet – Procedures for Atmospheric Testing in Confined Spaces, https://www.osha.gov/sites/default/files/publications/atmospheric_test_confined.pdf.

US Occupational Safety and Health Administration (OSHA), Hydrogen Sulfide, https://www.osha.gov/hydrogen-sulfide/hazards.

US United States Coast Guard (USCG), Issues Safety Alert on the Importance of Gas-free Certification and Compliance During Hot Work on Vessels, https://iimsusa.com/uscg-issues-safety-alert-on-the-importance-of-gas-free-certification-and-compliance-during-hot-work-on-vessels/#:~:text=%E2%80%93%20Occupational%20Safety%20and%20Health%20Administration,is%20safe%20for%20hot%20work.

US United States Coast Guard (USCG), Warning: Ensure a Gas-free Environment Prior to Conducting Hot Work, https://iimsusa.com/wp-content/uploads/2024/09/USCGSA_0324.pdf.

7

Controlling or Mitigation of OPE Hazards to Prevent Loss of Containment

Introduction

In this chapter, we will discuss in more detail how process equipment is prepared for isolation and the equipment and procedures required to ensure safe isolation for entry or hot work. Generally, the first step is to develop a plan for equipment isolation, including the development of a blind list, the types of blinds to be installed, and a layout, typically on the process and instrument drawings (P&ID) or an isometric drawing of the equipment to be isolated. The blind list becomes a controlled document, which is updated each time a blind is either installed or when one is removed. This is critical; we must be aware of the status of all blinds at all times.

Secondly, it is essential to determine where the isolation points will be located and how isolation will be achieved, preferably through double block and bleed valves to ensure 100% isolation. Once the primary isolation points (PIPs) are identified, we can determine where the opening process equipment (OPE) tags should be placed. This includes the isolation points, the energy verification, and the Open Here tags. These tags are important and are intended to prevent craftsmen from opening a piping system that has not been prepared for opening.

Proper equipment preparation is essential for process safety because it ensures a safe work environment for process technicians and mechanical craftsmen who open process equipment and perform maintenance while it is exposed to the atmosphere. This holds whether the equipment is opened for entry under OSHA-regulated confined space procedures or for hot work, such as burning and welding, on or inside the vessel while it is open. From my perspective, the equipment preparation process is essentially the same for confined space entry and for conducting hot work inside the vessel.

There is an exception to hot work on piping or other equipment while the equipment remains in service, which I will discuss separately later in this chapter.

The location of each equipment opening, including all blinds or blanks, should be marked on the P&ID and physically indicated on the pipe in the field. Marking the opening point, including for installing blinds, helps prevent workers from accidentally opening a flange or access plate on the wrong pipeline or vessel. This should be done using a uniquely designed or colored tag so workers can quickly recognize it as an opening or blind point. The tag should display the blind number that matches the blind list, the installation date, and the name of the person who installed the blind. This location will be confirmed during the on-site joint visit during the permit issuance process and at the time of the joint job site inspection. The joint visit will also verify that the worker performing the fieldwork accepts the work permit and knows the line opening location. As mentioned above, these locations must also match the blind list and the marked-up P&ID.

The equipment preparation procedure must specify the location of the PIPs, which may include double block and bleed valves – two valves in series with a bleeder between them to verify that the valves are sealing correctly and that no process fluid leaks through either the upstream or downstream isolation valve. Otherwise, especially if the equipment remains open for more than a few hours, equipment preparation generally involves using a blind to securely isolate the equipment. A blind is a blank plate with a tab protruding from the piping to indicate that it has been installed. Typically, a rated gasket is fitted on each side of the blind, but the side exposed to the process must definitely have a rated gasket. The tab should have a tag attached, indicating the blind number that matches the blind list, along with the installation date, and the name and contact information of the person who installed the blind.

The procedure should specify the process for safely draining the process fluid from the equipment and/or piping and how the material is safely transferred, handled, and stored. This should include the specifications for drain hoses, including considerations for design temperature, pressure, and for safely managing static electricity; typically, an all-bonded hose that is properly bonded and grounded should be specified. No hydrocarbon should be drained to the sewer system, and the procedure should include plans to dispose of all hazardous materials from the equipment draining operation properly.

There should be consideration for worker protection to chemicals used in the process to guard against skin contact or exposure to chemicals or chemical vapors. The procedure should specify personal protective equipment (PPE) based on the potential chemical hazards listed on the chemical Safety Data Sheet (SDS). This is typically goggles for eye protection, impervious gloves, and possibly chemical-resistant suits, depending on the chemicals that may be encountered.

Special consideration should be in place for any line or equipment openings that are planned for inside a confined space such as a building. Special precautions may be required due to potential confinement of spilled materials and for potential asphyxiation hazards, especially when nitrogen is used for purging.

The procedure should also specify the process for safely purging process equipment or piping of residual hydrocarbons or chemicals, including the purge medium to be used. This is typically done first with a medium boiling point, non-volatile hydrocarbon such as gas oil or diesel, followed by a steam purge or a sweep with nitrogen. The procedure should include safety precautions for working with steam or nitrogen, such as the potential for steam or condensate burns and the asphyxiation risks of nitrogen. The first two Wiley textbooks provide valuable guidance on each of these topics: *Process Operations Safety: The What, Why, and How Behind Safe Petrochemical Plant Operations* and *Hazards of Nitrogen and Other Inert Gases: How They Can Be Safely Managed.*

The procedure should highlight the potential for plugged bleeders and vent valves and specify how this should be handled using a fully encapsulated bleeder cleaning device if plugged bleeders are found. These devices can protect personnel while safely unplugging the small-bore valves used for drains and vents.

Operating Blinds

Some blinds are installed during equipment operation, such as to prevent sending a product to a disposition or storage tank not designed for it. For example, a blind in a liquefied petroleum gas (LPG) line can prevent accidentally sending LPG to an atmospheric storage tank, which is not designed to handle pressure. These are known as operating blinds or running blinds. These blinds should be labeled or tagged to prevent accidental removal during maintenance. Running blinds are typically removed during maintenance turnarounds to prepare the equipment for opening or to improve access. To ensure proper reinstallation, they must be recorded on the blind accountability list and signed off once reinstalled. When they are installed, labeling should identify the blind as an operating or running blind. See Figure 7.1 for an example of a "running blind" and its identification tag.

As mentioned above, when opening process equipment for entry or hot work, the steps are generally the same. The following outlines the key steps that the organization is responsible for when preparing process equipment for opening.

- During the preplanning phase, an isolation blind list should be created and thoroughly detailed with each isolation blind point.
  Note: There are two types of blinds. One is a flat metal plate with a tab, and the other is a flat metal plate with a tab and spacers attached to provide an air gap for physical separation and to facilitate a gas test.

Figure 7.1 Photograph of a label to identify an operations blind, also known as a "running blind."

- The blind list includes the physical location of each blind, its type, and the date and time it was installed. All blinds must be tracked to ensure that when a new blind is installed, it is added to the list, and when it is removed, it is deleted from the list. The list should always accurately reflect the blinds installed.
- Typically, the next step in equipment preparation is to flush it before opening. The process varies depending on the process unit, the metallurgy involved, and the type of material in the process. Usually, this involves switching to a light gas oil and flushing the equipment to a slop tank with the gas oil.
- The gas oil flush is typically followed by replacing the gas oil with water, then draining the water either into the process sewer or into a closed system for later treatment. For units with stainless steel piping or process vessels, like hydrofiners and hydrocrackers, the water must be tested for chlorides to prevent stress corrosion cracking. Due to the risk of catalyst damage and stress corrosion cracking in stainless steel, we usually use nitrogen to clear these piping circuits. Of course, when using nitrogen, we must be cautious of and prevent exposure to it due to the risk of asphyxiation from oxygen deficiency.
- After the water flush, the equipment is usually steamed to the atmosphere to eliminate any remaining hydrocarbons.

- Before opening process equipment, the potential for the presence of pyrophoric material must be considered and addressed. While this hazard will be discussed in more detail separately, it is essential to consider it, especially if the equipment has been in a higher hydrogen sulfide process. The primary requirement is to prevent the vessel interior from drying out when the vessel is opened. Periodic spray from a fire hose should be used until the vessel is fully opened and any deposits are safely removed.
- At this point, the process equipment can be opened to the atmosphere, and gas testing can be performed to ensure the equipment is free of flammable and toxic materials.

Exception for Hot Work on In-service Piping or Other Equipment

I mentioned earlier that there is an exception for hot work on piping or other equipment while the equipment remains in service. I will briefly cover this here since it rarely occurs, and when it does, it is carried out with special procedures and precautions.

Welding or performing other hot work on equipment that is in service is done occasionally; for example, welding a piping saddle onto an existing pipeline for a future connection or welding weld-plus ends after their installation. This is specialized work that requires careful planning, detailed procedures, and close attention to detail. The flow through the pipeline controls the heat generated during welding. The process flow is carefully monitored, and work is halted immediately if the flow is interrupted for any reason. Additionally, extensive testing of the piping system is conducted beforehand to ensure its integrity, including ultrasonic testing (UT) or x-ray inspection for piping thickness, as well as specialized welding procedures that use reduced heat and smaller welding passes, with multiple passes to ensure weld quality.

Equipment Preparation Procedure

The process or operations organization should be responsible for equipment and pipeline preparation before attempting to open any part of the system to the atmosphere. This includes developing a system isolation plan and procedure, preparing the unit or area blind list, which shows the location of each blind or blank, including blinds used to properly isolate the equipment and any "running blinds." Running blinds are typically installed to prevent sending products to an unsafe disposition that could cause process upsets or operating issues. These blinds are usually in place during unit operation, but must be removed when preparing equipment for opening. Of course, they must be reinstalled before the unit or equipment is recommissioned.

Once the equipment isolation envelope is identified, the main focus should be on ensuring that everyone involved in this project is fully committed to maintaining the integrity and level of the isolation. Therefore, the envelope should be as small and tight as possible, meaning that isolation valves or blinds should be placed as close to the planned work or entry point as possible. This approach offers several benefits, such as reducing the risk of gas or fluid leaks by minimizing the number of connections or cross-connections to other piping circuits or equipment. Additionally, the chance of trapped product is lowered because the volume within the tighter isolation envelope is smaller. It also helps prevent material from becoming trapped in stagnant piping or equipment.

The equipment isolation plan should specify the types of blinds to be used and where each blind will be installed. This advance planning is crucial to prevent trapped blinds and to ensure stand-off blinds are used for air circulation and gas testing locations on isolated vessels and piping systems, especially where entry or hot work is planned. The types of blinds, piping blanks, and their usage include the following:

- Flat plate isolation blinds, with a tab, sometimes called a spade blind;
- Figure 8 blinds, also known as spectacle blinds;
- Stand-off blinds with tabs attached to one side, sometimes known as spacer blinds;
- Bleeder blinds, sometimes known as vented blinds (with one side vented to a bleeder valve);
- Piping blank flange, including blank flange with bleeder valves attached.

Flat Plate Isolation Blinds (Also Known as Spade Blinds, or Slip Blinds)

These blinds are usually used to isolate equipment that is out of service for mechanical work. The flat plate isolation blind has a full gasket on the "live" side, providing separation from live piping and allowing mechanical work or inspection on the isolated side. These blinds should not be used for hot work on the isolated side; instead, stand-off blinds are used to create an air gap and

Figure 7.2 A flat plate isolation blind, sometimes also called a "spade blind" or a "slip blind."

facilitate gas testing. Additionally, these flat plate blinds are used to isolate piping or other equipment during hydrotesting. Please see Figure 7.2 for a photo of a flat plate isolation blind.

Sometimes, there can be confusion between a bleeder blind and a flat plate blind. Many bleeder blinds feature a notched plate that allows one side of the blind to be vented. These usually have a "V" notch in the handle to help identify them as bleeder blinds rather than flat plate blinds. Bleeder blinds are also explained in more detail below.

Figure 8 Blind, Also Called a "Spectacle Blind"

Figure 8 blinds function like flat plate blinds; however, these blinds have an open end and a closed end and can be rotated to either the open or closed position. For convenience, they are used in locations where frequent blinding is needed. Examples of Figure 8 blinds are illustrated in Figures 7.3 and 7.4.

Figure 7.3 Image of a Figure 8 or Spectacle blind.

Figure 7.4 Photograph of two Figure 8 blinds, installed in the open position.

Stand-off Blinds

Stand-off blinds with stand-off tabs attached on one side are commonly used when isolating process vessels or preparing for hot work on the isolated section of pipe. The stand-off tabs allow air circulation into the vessel or pipe and facilitate gas testing.

When isolating process vessels such as fractionation columns, storage tanks, and other process vessels, and/or piping where hot work is planned, a blind with stand-off tabs attached should be installed with the tabs against the vessel being opened. This provides physical separation and an air gap between the process vessel and piping systems. We always referred to these blinds as "stand-off blinds" because of the tab that separates the vessel from the piping or valve. This gas allows air to circulate into the process vessel and provides a place where gas testing can be done.

Please see Figures 7.5 and 7.6 for simplified sketches illustrating the stand-off blinds and how they provide a physical separation from the process piping systems.

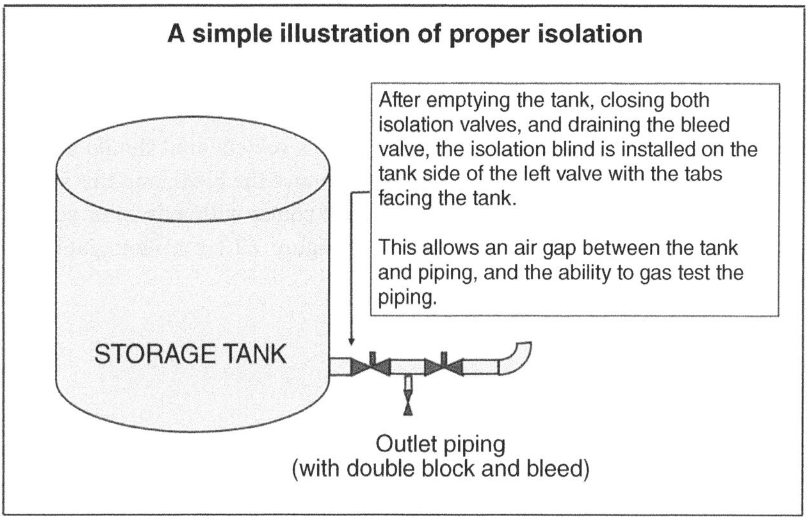

Figure 7.5 An example of where a stand-off blind may be used to provide an air gap separation and help facilitate a gas test.

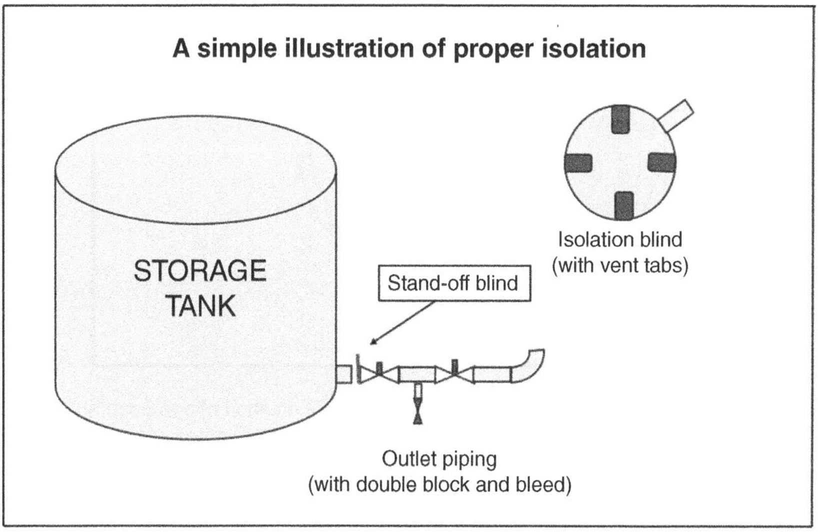

Figure 7.6 An example of the use of a stand-off blind.

Bleeder Blinds, Also Called a "Vented Blind"

Bleeder blinds are built with one side vented to a bleeder valve. Be cautious of unintentionally creating a trapped blind, where one side might become pressurized without a way to vent during removal. Ensure each blind location includes a bleeder, vent, drain, or other means of verifying zero energy on both sides of the blind. Otherwise, process fluids or pressure can become trapped behind the blind, potentially causing a release when removing it.

If installing a blind results in a "trapped blind," a vented blind should be used to help drain trapped liquids when it's time to remove the blind, and this should be considered before installation. A vented blind comes with a drain or vent on the pressurized side once installed. Please see Figure 7.7 for a photograph of a vented blind.

Piping Blank Flange

This is the most secure method of process isolation. The pipe is completely disconnected, and the blank flange is attached to the remaining pipe section. This creates an air gap between the piping and the vessel or piping where the work is being done. Figure 7.8 shows two installed piping blank flanges.

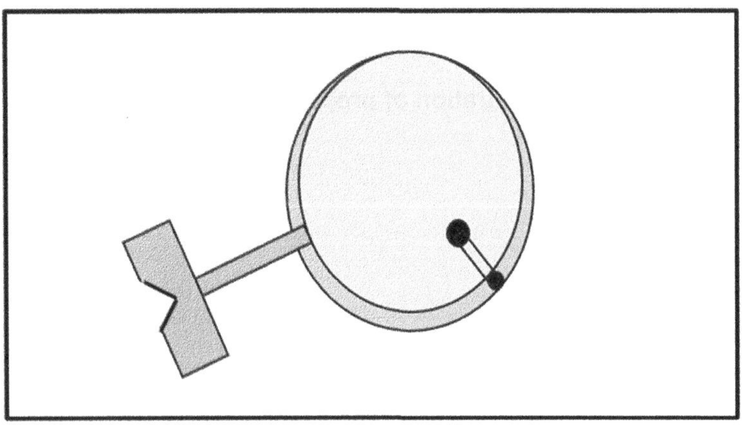

Figure 7.7 Image of a vented blind, used to prevent the creation of a trapped blind, with accumulation of pressure on one side that cannot otherwise be vented when the blind is to be removed. This blind is vented on one side to prevent pressure accumulation, which may occur due to leaking isolation valves. Vented blinds may also be fitted with a bleeder valve in the vent port, which is typically referred to as a "bleeder blind."

Figure 7.8 Installed blank flanges on piping systems.

Proper System Isolation

There have been many examples of process safety incidents, including fatalities, resulting from attempts to isolate parts of a unit to allow work on process equipment while the rest of the unit remains operational. For instance, two incidents in Chapter 11, "What Has Happened/What Can Happen," fall into this category. At the Tosco Avon Refinery, in February 1999, operations continued at the crude distillation unit while contractors attempted to replace a section of the overhead naphtha line. Unfortunately, during this operation, a release of light naphtha occurred, quickly followed by a major fire that killed four workers and caused a life-changing injury to a fifth worker. Additionally, in June 2001, work was underway at the Chevron Pembroke (UK) Refinery to isolate and prepare a storage tank on the amine unit for cleaning and maintenance. The refinery recognized that the experienced work crew would not be available, and a suggestion was made to delay this work until the experienced crew returned. However, the less experienced crew attempted to isolate and prepare the tank, resulting in a catastrophic explosion that killed four workers and caused a life-altering injury to a fifth person.

To prevent such tragedies, equipment isolation should only be performed by experienced craftsmen under close oversight by the operations team. Our experience shows that it is generally unwise to isolate a small part of the unit while the rest continues to operate. Operation risks require a thorough evaluation to assess the danger and consider options such as postponing the work to a scheduled turnaround or a safer time when experienced crews are available, or when the inventory of hazardous materials is lower. An evaluation might reduce risk by delaying work on piping systems still connected to equipment that remains in operation.

When isolating multiple process vessels or other equipment, it is also not recommended to attempt to blind or otherwise isolate two or more vessels or pieces of equipment as a single isolated "circuit" by isolating them together as one vessel. In such cases, each process vessel or piece of equipment should be isolated individually, blinded, isolated, or checked as a single vessel, and gas tested for opening and entry as separate vessels or pieces of process equipment.

A thorough walkdown of the system should be conducted by the responsible parties to confirm that the system to be opened is completely isolated from all other in-service or pressurized equipment. All valves on the main system must be fully closed, including all valves on system interconnections or cross connections between the system being isolated and any other system. All bleeders and vents should be fully open and checked to ensure they are free of obstruction.

This process should create an isolated envelope around the system to be opened, with no other cross connections unless they are also fully isolated.

Control valves should never be used as an isolation device. They are not reliable, and excess pressure under the seat may result in the valve moving to a fully open position. Also, there is always the possibility that the valve activation medium, instrument air, electrical power, or hydraulics may inadvertently be applied (or in some cases, released), causing the valve to open. For air-to-close control valves, the valve will travel fully open if instrument air is lost.

If at any point during the isolation process or when opening the line, there is any sign that the isolation valves are leaking, all work must stop immediately until it is confirmed that isolation has been restored and the system is at zero energy.

Ensure Overpressure Protection

Be careful of blocking in piping systems or equipment, especially until vents and drains are opened to prevent inadvertent pressure buildup in the isolated system. This is particularly important if any heat may be applied to the isolated system, including heat from atmospheric temperature increase or exposure to the sun. Until vents and drains have been secured in the open position, it is a good idea to ensure the relief valve(s) remain in service which can relieve pressure in the system.

OPE Area Barricading and Signage

It is important to set up barricades around the area where isolation blinds are being installed or process equipment is being opened. These barricades should be reinforced with warning signs to alert personnel who are not involved in the OPE process of the ongoing work and the potential risks of injury or contact with process materials.

OPE "Open Here" Flange/Access Plate Unique Identification Tags

In this book, I often mention using a unique tag to identify blind points or other flange or access plate openings. This paragraph offers more details about the purpose and use of these "unique" tags and how they can help ensure that only the intended flanges are opened. In other words, this unique tag's design and application prevent the accidental opening of the wrong flange or access plates.

Tag design: The OPE "Open Here" tag should be of a unique color and design and should only be used for this purpose and no other purpose. It should be readily identifiable to the field personnel to indicate that the tagged flange or plate has been prepared and is ready to be opened for blinding or vessel opening. At our site, we referred to these tags as "crazy tags" to ensure that the tag was unique and used for only this purpose. It was colored in psychedelic colors and clearly labeled as an OPE "Open Here" tag, so it could not be confused with any other tag.

Who installs the OPE "Open Here" tag? The operations group is responsible for placing the OPE "Open Here" tags on equipment opening points. This ensures that tags are not inadvertently installed on the wrong equipment and are only installed after the equipment has been prepared for opening.

When is this tag applied? The OPE "Open Here" tag is installed only after the piping system or equipment has been fully prepared for opening. This means that the equipment has been drained, flushed or purged, is in a "zero-energy" state, and is ready to be opened. This ensures that equipment is not inadvertently opened before preparation is completed.

How is the OPE "Open Here" tag affixed? The design of this tag should include a tamperproof and weatherproof attachment to affix it to a flange or access plate. The lettering should also be weatherproof to ensure readability, typically with a permanent marker or similar.

Important Consideration

> *The OPE "Open Here" tag is designed to prevent the craftsmen from inadvertently opening the wrong flange or access plate. They MUST be affixed to each flange or access plate to be opened or removed ONLY after the equipment is prepared for opening.*
>
> *The craftsmen MUST understand that they are not to open any flange or access plate unless this tag is attached.*

Lockout/Tagout Procedures

The procedure should also ensure that equipment complies with the lockout provisions of OSHA 29 CFR 1910.147. All energy sources must be isolated, and PIPs should be identified in the field and on the accompanying process P&ID (instrumentation and design drawing) before any attempt to open the equipment. The PIP points should be consistent with those specified in the equipment preparation procedure.

The PIP points, such as isolation valves, vents, drains on process piping and equipment, electrical breakers, or other forms of energy isolation, should be identified with a PIP tag.

Joint Job Site Visit (Between the Work Permit Issuer and the Person(s) Who Will Be Opening the Flanges)

The joint job site visit is a meeting held at the job site between the work permit issuer, who authorizes the work to begin, and the tradesmen or craftsmen (the workers) who will be doing the work, or their representative. Communication is always clearer if the permit issuer can meet directly with the workers, but unfortunately, this is not always possible. If the issuer meets with their representative, that person is responsible for asking any relevant questions on behalf of the workers and communicating all aspects of the work to be done to them.

The importance of an effective joint job site visit cannot be overstated. This is the last chance to ensure everything is correct, including a clear understanding of which equipment will be opened, the procedures or techniques to be used, and any special considerations or limitations that must be followed. Clear communication is crucial; if there are any questions or uncertainties, this is the time to resolve them. For example, as mentioned earlier, hot work should never be approved for OPE, and no hot work should be performed near flanges that are being opened. During the joint site visit discussion, this should be reaffirmed between the permit issuer and the workers.

During the joint job site visit, a walk-down is conducted between the permit issuer and the workers, and the workers should identify every flange and/or access plate that needs to be opened or removed. This ensures clear communication and confirms that the workers are familiar with the work to be completed.

This review should include detailed discussions of the energy isolation process and demonstrate that all equipment is in a "zero-energy" state. This includes the following:

- Equipment is drained and depressured;
- All forms of energy are isolated and discharged, including the following:
  - Process pressure
  - Electrical energy
  - Spring energy
  - Radioactive energy or sources
  - Gravitational energy
  - All other energy forms or sources
- Equipment is purged and free of hydrocarbons, chemicals, or other toxic or flammable materials;
- During equipment opening, all residual materials are controlled or captured;
- Personnel are protected by the proper use of PPE or other safeguards and by defensive measures.

The discussion should also confirm that no bolts shall be completely removed from a flange or access plate until all bolts have been loosened and workers have verified that there is no pressure or fluid leakage from the flange being opened. Following this procedure, the workers have the ability to reclose the flange if product or pressure is present.

If special PPE is required to protect the workers from process hazards associated with any chemical residue remaining in the equipment, this should be included on the work permit and discussed during the joint job site visit. This includes goggles, protective clothing, self-contained breathing apparatus (SCBA), etc.

Likewise, any special fire protection equipment should be included on the work permit and discussed with the workers during the joint job site visit. For example, when opening fractionators or storage tanks, especially those that have been in sour service (high concentrations of hydrogen sulfide), there is the possibility of pyrophoric iron sulfide, which can ignite remaining hydrocarbons or residues and can even ignite metal demister screens or metal tower packing, which burns exceptionally hot and is extremely hard to extinguish. For those cases, it may be appropriate to specify on the work permit that a charged fire hose be available at each manway as they are removed. It is a good idea that, for these cases, requiring periodic wetting of the vessel internals helps to prevent the pyrophoric material from drying, which can result in ignition. In these cases, this should be specified on the work permit and included in the joint job site discussion.

Precautions When Opening Flanges

Process personnel should verify that all pipelines and other process equipment to be opened is properly drained and free of hazardous material. This includes verification that all drains, bleeders, and vents are fully open and free and clear. Ensure awareness that bleeders or vents may be obstructed; therefore, it is necessary to reaffirm that no plugging or obstructions exist. Please remember that we should handle disconnecting a blind flange or a blank as a line opening.

All equipment to be opened must be verified to ensure zero energy in the system(s). If zero energy cannot be verified, a risk assessment must be performed, and alternate ways to demonstrate zero energy must be implemented. This may involve special procedures to drill into the piping with a small-diameter pneumatic drill while the workers are wearing appropriate PPE to protect them from the process fluids that were previously contained. There should be ready access to a clamp or other closure device in the event of pressure or fluids being identified.

If the line contains or has contained flammable material and normal line opening fails, it may be necessary to cold-cut the piping. If this is the case, nitrogen purge should be considered to help reduce the potential fire or explosion hazards. When using nitrogen, always be aware of the asphyxiation hazards and be careful to vent in areas where there are no personnel present. Also, when cold cutting pipe be aware of the static hazards and provide grounding straps on both sides of the pipe to be cut.

All workers, company personnel or contract employees, should be trained in the proper procedures and practices associated with the opening of process equipment. The following provides general guidance, which should be converted to site guideline or policy and communicated to all workers before work is started:

- A preliminary gas test should be conducted at the bleeders and vents with tests for oxygen (O_2), flammables (% LEL), hydrogen sulfide (H_2S), and other chemicals, which may be present depending on the service. This preliminary gas test is used to determine the PPE to be prescribed for the workers during the OPE operation. Typical pass/fail criteria are listed below. This is only guidance; please be sure to follow regulatory requirements and your company's procedure.
  - Oxygen content: 19.5% minimum to 23.5% maximum
  - Flammables: 10% LEL maximum
  - Hydrogen sulfide: 20 ppm maximum

- Prescribe the PPE for the workers based on the preliminary gas test plus protective wear such as gloves, goggles, chemical respirators, or SCBA in the event of chemical exposure. All PPE must be continued until the system is fully open and determined to be free of hydrocarbons and process chemicals.
- Workers should be trained to never remove all the bolts or nuts from the flange or access plate before attempting to break the seal. Instead, they should loosen the bolts and break the flange seating (the seal) while the bolts are loose but still in place. This allows the workers to quickly reclose the flange if pressure or fluids are discovered when the seal is broken.
- Workers should be trained to remain out of the line of fire when opening flanges and to loosen the flange bolts away from them in the event of trapped energy.
- Hot work should never be used to burn bolts while opening flanges on piping or other process equipment. It is acceptable to "hot bolt" flanges if the area is confirmed to be free of flammables. Hot bolting is described as burning one bolt at a time, replacing the bolt, and tightening it before burning the next bolt. This process is continued until all the bolts are replaced, and then the flange is opened the normal way, without hot work.
- If any leakage is detected during the flange opening process, the work must immediately be ceased and a path forward must be discussed with the lead process individual.
- No hot work, flame or non-flame, should be permitted in process areas where equipment and piping are being opened. This means no burning or welding, no arc gouging, and no gasoline or diesel-powered vehicles or engines of any kind. These are ignition sources, and any released fuel may result in an explosion or major fire.

Confined Space Entry

Many workers have been killed by entering a confined space without following the confined space entry standard (OSHA 29 CFR 1910.146). It only takes one deviation or omission, and one or more people can die. A link to the OSHA standard is provided in the Additional References section of this Chapter, and you are encouraged to print the standard and become thoroughly familiar with the requirements. If you notice any deviations, steps, or requirements not being followed in the field, I encourage you to speak up and get the workers back on track. Your actions can save lives!

A common issue I've encountered in my career is that, immediately after opening a process vessel with a history of operational problems, workers and even managers often want to look inside to find the cause of the issue or understand why the process isn't working as expected. This is not allowed under OSHA standards, which define vessel entry as any part of a person's body being inside the vessel. All requirements of the confined space standard must be fulfilled, including having a fully completed confined space entry permit before entry. It is our role to ensure that the confined space standard is enforced in its entirety.

Gas testing prior to entry is one of the most important considerations to protect life and health. The gas test should always be done by a qualified individual and with a properly calibrated and bump-tested gas test machine. Bump testing verifies the accuracy of the gas test instrument by testing a sample with a known concentration of flammable gas, hydrogen sulfide, or the specific chemical being sampled for. When sampling a vessel, several tests should be done at various levels throughout the vessel to verify that there is no gas stratification. There have been cases where the vessel was tested at or near the entry point, only to have an explosion due to pockets of gas in other parts of the vessel.

For a moment, I'll go back to a point made in an earlier chapter. If the vessel being tested contains a combustible liquid at ambient temperature (less than 100 °F [38 °C]), the gas test machine will not read vapors. Assuming no other gas is present, the machine will read 0 LEL. For example, in a vessel containing diesel or gas oil with a flash point well above 100 °F, the liquid will not vaporize at ambient temperature, and no vapor will be present. However, if the work to be done involves hot work (burning or welding), the heat can quickly heat and vaporize the diesel, resulting in the potential for a significant fire inside the vessel. Please be aware that 0 LEL does not mean that the vessel is hydrocarbon-free. It simply means that no vapors were present at the time the gas test was completed.

Normally, the vessel entry should be done within two hours of the gas test. This is another important consideration, as the heat of the day can vaporize hydrocarbons from sludge or from scale on the vessel internals, and as a result, explosions have occurred, and people have been killed. One way to avoid this is by using continuous gas testing. I am a huge fan of continuous gas testing. With continuous testing, the gas testing is done in real time, and the results are instantaneous. The price and availability of continuous gas testing machines have come into range where there is little reason why continuous gas testing is not more widely used.

A summary of the primary requirements of the OSHA confined space entry standard is summarized below (note: this is only a summary, please refer to the entire standard for additional details):

- Permit-required confined space (permit space) means a confined space that has one or more of the following characteristics:
  1. Contains or has a potential to contain a hazardous atmosphere;
  2. Contains a material that has the potential for engulfing an entrant;
  3. Has an internal configuration such that an entrant could be trapped or asphyxiated by inwardly converging walls or by a floor which slopes downward and tapers to a smaller cross-section; or

4. Contains any other recognized serious safety or health hazard.

- If the workplace contains permit spaces, the employer shall inform exposed employees, by posting danger signs or by any other equally effective means, of the existence and location of and the danger posed by the permit spaces.
  Note: A sign reading "DANGER – PERMIT-REQUIRED CONFINED SPACE, DO NOT ENTER" or using other similar language would satisfy the requirement for a sign.
- If the employer decides that its employees will enter permit spaces, the employer shall develop and implement a written permit space program that complies with this section. The written program shall be available for inspection by employees and their authorized representatives.
- When entrance covers are removed, the opening shall be promptly guarded by a railing, temporary cover, or other temporary barrier that will prevent an accidental fall through the opening and that will protect each employee working in the space from foreign objects entering the space.
- Before an employee enters the space, the internal atmosphere shall be tested, with a calibrated direct-reading instrument, for oxygen content, for flammable gases and vapors, and for potential toxic air contaminants, in that order. Any employee who enters the space, or that employee's authorized representative, shall be provided an opportunity to observe the pre-entry testing required by this paragraph.
- There may be no hazardous atmosphere within the space whenever any employee is inside the space.
- Continuous forced air ventilation shall be used as follows:
  - An employee may not enter the space until the forced air ventilation has eliminated any hazardous atmosphere.
  - The forced air ventilation shall be directed to ventilate the immediate areas where an employee is or will be present within the space and shall continue until all employees have left the space.
  - The air supply for the forced air ventilation shall be from a clean source and may not increase the hazards in space.
- The atmosphere within the space shall be periodically tested as necessary to ensure that continuous forced air ventilation prevents the accumulation of a hazardous atmosphere. Any employee who enters the space or that employee's authorized representative shall be provided with an opportunity to observe the periodic testing required by this paragraph.

- Provide at least one attendant outside the permit space into which entry is authorized for the duration of entry operations:
  - Note: Attendants may be assigned to monitor more than one permit space, provided the duties described in paragraph (i) of this section can be effectively performed for each monitored permit space. Likewise, attendants may be stationed at any location outside the permit space to be monitored as long as the duties described in paragraph (i) of this section can be effectively performed for each monitored permit space.
  - If multiple spaces are to be monitored by a single attendant, include in the permit program the means and procedures to enable the attendant to respond to an emergency affecting one or more of the permit spaces being monitored without distraction from the attendant's responsibilities under paragraph (i) of this section.
- Develop and implement procedures for summoning rescue and emergency services, for rescuing entrants from permit spaces, for providing necessary emergency services to rescued employees, and for preventing unauthorized personnel from attempting a rescue.
- The employer shall retain each canceled entry permit for at least 1 year to facilitate the review of the permit-required confined space program required by paragraph (d)(14) of this section. Any problems encountered during an entry operation shall be noted on the pertinent permit so that appropriate revisions to the permit space program can be made.

End-of-Chapter Quiz

1 At what point in the equipment isolation and preparation phase should an isolation blind list be created, thoroughly detailing each isolation blind point? Please select the correct answer.

a During the preplanning phase.
b After the unit shutdown, before field work starts.
c It's okay to make the list as blinds are being installed.
d As an integral part of the overall turnaround plan.

2 Please explain why three different types of equipment blinds are used and the primary difference between them.

3 Who participates in the joint job site visit, and what is its purpose?

4 Why must equipment isolation blinds be placed as near the isolated equipment as possible?

5 What training should workers receive relative to opening flanges on process piping?

6 What type of hot work is allowed in proximity to process equipment that is being opened for isolation blinding? Why is your answer important?

7 Many workers have been ________ by entering a confined space without following the confined space entry standard (OSHA 29 CFR 1910.146). It only takes _______ deviation or omission, and one or more people can die.

8 How long is the employer required to retain the completed permit-required entry work permit?

9 The US OSHA Confined Space Entry Standard requires that before an employee enters the space, the internal atmosphere shall be _________, with a calibrated direct-reading instrument, for ________ content, for __________ gases and vapors, and for potential __________ air contaminants, in that order.

10 Please describe what is defined as a "trapped" blind?

11 What is the hazard with a trapped blind?

12 When preparing process equipment for isolation or opening, all forms of energy must be eliminated. The primary energy source is process pressure. What are other forms of energy that must also be eliminated?

13 Process personnel should verify that all pipelines and other process equipment to be opened are properly _________ and free of __________ material.

14 This includes verifying that all drains, bleeders, and vents are fully open, free and clear. Ensure awareness that bleeders or vents may be________; therefore, it is necessary to reaffirm that no plugging or obstructions exist.

15 Please remember that we should handle disconnecting a blind flange or a blank as a _______ __________.

16 Is it acceptable to burn bolts to open process equipment?

17 Describe what is meant by "hot bolting."

18 What are the significant benefits of continuous gas testing compared to the daily gas test?

19 What is the purpose of the opening process equipment (OPE) tag, and how is it used?

20 Our experience shows that it is generally not wise to _______ a small section of the unit while the rest of the unit continues to operate. Operation risks require a thorough evaluation to assess the _______ and consider options such as postponing the work to a scheduled turnaround, a safer time when experienced crews are available, or when the inventory of hazardous materials is lower. An evaluation might reduce ________ by delaying work on piping systems still connected to equipment that is in operation.

21 Following the guidance in this book, the OPE "Open Here" tag is installed on the specified pipeline and vessel flanges.
Please select all that apply.
- **a** There is no guidance for "Open Here" tags on pipeline flanges or vessel access plates.
- **b** Installing an "Open Here" tag on every other flange would be acceptable because workers can see them.
- **c** "Open Here" tags are used solely on pipeline flanges, not on process vessel access plates.
- **d** "Open Here" tags should be attached to each pipeline flange and process vessel access plate that will be opened or removed.

22 What is the purpose of the OPE "Open Here" tags, and why is this important?

Additional References

American Institute of Chemical Engineers (AICHE), Confined Space Entry – Strategies & Effective Practices to Manage and Mitigate Hazards, https://www.aiche.org/ccps/resources/tools/safe-work-practices/confined-space-entry/strategies-effective-practices-manage-and-mitigate-hazards.

American Institute of Chemical Engineers (AICHE), Line Opening – Strategies & Effective Practices to Manage and Mitigate Hazards, https://www.aiche.org/ccps/resources/tools/safe-work-practices/line-opening/strategies-effective-practices-manage-and-mitigate-hazards.

ExxonMobil, Opening Process Equipment – Important Considerations, https://refiningcommunity.com/wp-content/uploads/2017/07/Opening-Process-Equipment-OPE-Important-Considerations-Moloney-ExxonMobil-DCU-FCCU-SRU-Budapest-2017.pdf.

US Chemical Safety and Hazard Investigation Board (CSB), Final DCRC Bulletin, https://www.csb.gov/delaware-city-refining-company/.

US Chemical Safety and Hazard Investigation Board (CSB), Safety Digest: CSB Investigations of Incidents during Startups and Shutdowns, https://www.csb.gov/assets/1/6/csb_digest_-_startup_shutdown.pdf.

Westlake Chemical Corporation; Lake Charles, Safe Process Piping, Equipment & Vessel Opening Procedure, https://www.westlake.com/sites/default/files/Safe%20Process%20Piping,%20Equipment%20and%20Vessel%20Opening%20Procedure.pdf.

8

Lifesaving Rules and the Consequences of Violating Lifesaving Rules

Lifesaving Rules

Many facilities have what they call "lifesaving rules," "critical safety rules," or other special designations. These typically serve as safety rules aimed at preventing situations where a single unsafe act, failure to follow a critical procedure, or similar deliberate errors could lead to severe injury, major loss of containment, or even fatalities.

These are usually a small set of critical rules designed to save lives. Full compliance with the lifesaving rules should be mandatory and a condition of employment. In other words, failure to follow these rules can result in immediate termination. This may seem strict, rigid, or even overly harsh, but remember, these rules are meant to protect workers' lives. I believe, and my experience confirms, that once these rules are in place for a while, employees will understand their purpose and realize they are there to safeguard their lives. Of course, the success of Lifesaving rules in any company depends on strong commitment from management and support from employees.

I fully support the implementation of lifesaving rules at each manufacturing site, and this practice has proven to reduce the number of serious incidents, especially those that can lead to or have resulted in catastrophic injuries or loss of life. An example of a lifesaving Rule is the failure to wear fall protection when working on a platform or scaffold at an upper elevation, particularly when the platform or scaffold is not fully equipped with appropriate handrails, guardrails, or protected (caged) ladders. In this situation, it's easy to see how a slip or fall could result in a serious injury or death. When properly used, fall protection such as a harness and lanyard, or a harness connected to a retractable lanyard with a quick-action braking system, can save lives. My point is that working on an incomplete scaffold without this type of fall protection should be considered a violation of the lifesaving rules, which is a serious offense.

I also support implementing this policy at all facilities where high concentrations of H_2S are present or could be present, requiring personnel to wear an H_2S detector whenever they are in areas with operating equipment containing elevated hydrogen sulfide levels. I believe this policy should be officially adopted by the facility and enforced as a lifesaving rule. This means that wearing the detector in a high H_2S area would be a condition of continued employment at the facility. Most affected industries, such as oil and gas and municipal sewer plants, should find it straightforward to enforce this through their existing internal organizational structures.

In my previous textbook on the *Hydrogen Sulfide (H_2S) Hazards: Risk Recognition and Prevention*, I discussed two incidents where the operator was assigned an H_2S detector but was not wearing it when exposed. In both cases, it cost the operator their life; in one, it also cost the life of the operator's wife. Wearing an H_2S detector should be like putting on your seat belt – it should be automatic and require no thought. This is what can happen when these are enforced as lifesaving rules. Table 8.1 provides several examples of such rules.

The following explains the reasoning behind the current list of lifesaving rules and why these rules are crucial. As you'll see, the consequences in each case are catastrophic for the person involved.

Table 8.1 Examples of lifesaving rules and their rationale.

1. Always use approved fall protection when working at heights on an unprotected platform or scaffold.
   Approved fall protection should be worn anytime workers are on an incomplete platform, incomplete scaffold, or uncaged ladder. The fall protection is designed to protect workers from an unprotected fall that could result in severe injury or death.
2. Obtain an approved hot work permit before performing hot work.
   The hot work permit details the job site prerequisites, sewer protection, fire protection, gas testing, and other task requirements, and is the authorization by the responsible party for hot work (burning and welding) to begin.
3. Always perform a gas test, have a standby attendant, and obtain proper authorization before entering a confined space.
   Entering a confined space must meet the detailed requirements of the OSHA confined space regulation, including detailed gas testing, outside attendant, fresh air ventilation, emergency rescue plan, and entry authorization (the entry permit by the authorized person). Entering a confined space without meeting these requirements can and has resulted in death to the entrants.
4. Obtain authorization and ensure proper mitigation before disabling a safety-critical device.
   Safety-critical devices are the last line of defense against loss of containment, which can lead to explosion or fire. Before a safety-critical device can be disabled, there must be mitigation in place that replaces the functionality of the safety-critical device being defeated. The defeat must be authorized by the approving authority for a specified period of time, and the defeat must be closed after the device is returned to full online service with all bypasses and defeats removed.

5. Always verify proper equipment isolation before beginning opening process equipment (OPE) work.
   Failure to determine zero energy has resulted in the loss of containment of flammable or toxic materials, fires and explosions, and major disruption to the surrounding environment and communities. Failure to determine zero energy has also resulted in severe injury and death to workers.
6. Always wear your seat belt when driving or riding a vehicle.
   In the event of a collision, the failure to wear a seat belt has resulted in ejection from the vehicle and/or the resulting severe or fatal injury to the driver and passengers.
7. No use of alcohol or drugs, and none in your system, while working or driving.
   Alcohol and drug use results in mental and physical impairment and leads to operating mistakes and poor decisions. Alcohol and drug use is also a stated violation of company policy in addition to a violation of the lifesaving rules.
8. No smoking, except in clearly marked designated areas.
   Smoking is a potential source of ignition for most hydrocarbons and is, therefore, only allowed in designated safe areas where the potential for a hydrocarbon release is minimized. Smoking is never allowed in areas where hydrocarbons or other chemicals may be present.
9. Always wear a personal H_2S detector when in H_2S areas.
   A personal H_2S detector will provide an early warning of the presence of hydrogen sulfide. H_2S is noticeable at extremely low concentrations; however, the deadly gas deadens the olfactory nerves at about 100–150 ppm, causing the wearer to believe the gas has dissipated or gone away. The detector will continue to alarm even though the wearer can no longer detect H_2S by odor. The wearer should be prepared to immediately escape from the area anytime the alarm sounds.
10. Never walk or place any part of the body under a suspended load.
    Depending on the crane, load lines, or other rigging, a suspended load can be an extreme hazard. If any part of it fails, it can quickly split, becoming a series of dangerous and deadly projectiles. Always keep all parts of the body completely clear of suspended loads.

Restated Examples of Lifesaving Rules

1. Failure to use fall protection when working at unprotected heights. For example, on incomplete scaffolds, platforms, uncaged ladders, pipe rack or unit structure.
2. Failure to obtain a hot work permit before burning or cutting.
3. Entering a confined space entry without authorization (vessel, or excavation >4 feet deep).
4. Disabling a safety-critical device without authorization.
5. Failure to verify equipment isolation before beginning OPE work.
6. Failure to wear a seat belt when driving or riding in a vehicle.

7. Alcohol or drugs found in the worker's system while working or driving.
8. Smoking in areas not designated as authorized smoking areas.
9. Failure to wear a personal H_2S detector when in H_2S areas.
10. Walking or placing any part of the body under a suspended load.

Note: These are only examples. This is not intended to be a complete list.

End-of-Chapter Quiz

1 Please describe in your own words what the term "lifesaving rules" means.

2 Regarding a lifesaving rule, if one of these rules is violated, what is a possible consequence of the violation? Why should this be a possible consequence?

3 Full compliance with the lifesaving rules should be considered __________ and a condition of ___________.

4 When working in high-concentration H_2S areas, putting on an H_2S detector should be like putting on your seat belt; you shouldn't have to think about it.

True or False

5 The practice of requiring lifesaving rules has proven to reduce the number of ________ incidents, especially those that can result in, or have resulted in, catastrophic ________ or loss of _______.

6 Please list as many of the potential lifesaving rules as possible.

Additional References

Bourbon Off Shore, The 12 Life Saving Rules, https://www.bourbonoffshore.com/sites/default/files/2023-05/Leaflet%20LSR%20-%20VA%20pour%20t%C3%A91%C3%A9chargement%20site%20web.pdf.

ConocoPhillips Safety, Health, and Environment – ConocoPhillips Life Saving Rules, https://static.conocophillips.com/files/resources/life-saving-rules-minimum-standards2020.pdf.

Energy Safety Canada, Life Saving Rules, https://www.energysafetycanada.com/Standards/Industry_Standards/Life-Saving-Rules.

International Association of Oil and Gas Producers, Life-Saving Rules – IOGP Report 459, https://www.iogp.org/bookstore/product/life-saving-rules/.

9

The Hazards of Risk Perception and Risk Habituation

Risk Perception vs. Risk Habituation

First, let's make sure we all agree on risk perception and risk habituation. risk perception is how we each judge the level of danger from a specific hazard, and this judgment depends on many factors. These include our background and experience, training, and a lifetime of decision-making, some of which are good, while others are less so. Our perception of risk directly influences our behavior. For example, when driving to and from work, we should consider the likelihood of being involved in a serious traffic accident. If we truly considered this, I believe it would help ensure that we consistently use our seat belts.

Risk habituation (RH) is a psychological process where repeated exposure to a certain hazard without consequences can lead to a decreased perception of risk and may cause people to take shortcuts, bypass procedures, and make other decisions that can result in accidents and injuries.

Risk perception refers to how each person assesses a specific risk, while RH is the decrease in our sensitivity to that risk. Studies have shown that when we are repeatedly exposed to high-risk work, we tend to lose focus on the actual hazards involved. This phenomenon, known as risk habituation, is sometimes called complacency or becoming complacent at work. It can lead to taking risks for granted and engaging in behaviors that increase our danger while ignoring potential consequences. However, there is more to it, and that is what we will explore in this brief chapter.

The National Institute of Health (NIH), National Library of Medicine, and National Center for Biotechnology Information reported in an article titled "Risk Perception: It's Personal" the following: "Risk perception is a highly personal process of decision making, based on an individual's frame of reference developed over a lifetime, among many other factors. A body of research from the past several decades makes it clear that when it comes to making decisions about health and safety, we don't always worry the most about the most pressing threats." NIH also reported that RH directly affects our risk perception, and our individual assessment of an ongoing risk or hazard can be reduced. This frequently occurs when we are repeatedly exposed to the potential for serious injury, but no one is harmed.

Examples Where RH Has Occurred

Highway Construction Work

Roadwork is a high-risk job because workers are at risk of being hit by oncoming traffic. You're probably aware of this or have at least heard of workers being struck by vehicles and sustaining serious injuries or even losing their lives. Since workers are often working around fast-moving vehicles, they face significant danger. Numerous studies have been conducted to address this hazardous issue. You may have seen some of their results in today's road construction projects. Measures such as highly visible road barriers, reduced speed limits in construction zones, and parking trucks or other vehicles with impact-resistant bumpers help mitigate these risks. Most U.S. states have increased speeding penalties for speeding in construction zones.

There have also been programs aimed at addressing the physiological factors that contribute to reducing workers' stress and fatigue, as these can directly impact decision-making. Additionally, workers' mental health issues, like depression or anxiety, can be linked to unsafe behaviors. Other factors identified include high job demands, time pressure, and a lack of support from management or supervision. Poor working conditions, such as long work hours and job dissatisfaction, also play a role and are being addressed by the industry.

Falls from Escalators in Offices and Public Buildings

I understand that falls from escalators have happened, but I was almost stunned by the number of articles written by legal firms on this topic when I looked for the data. It seems this occurs more often than I had thought. Falls from escalators can cause serious and disabling injuries, and some have even resulted in death. As I reviewed the data, I found that falls from escalators have caused broken bones, including broken legs, arms, backs, hips, and more. Many of these also involved concussions and other brain and neck injuries.

I believe many of these incidents happen because of how we perceive risks, possibly due to RH. After all, escalators are a normal part of our lives, and when used correctly, they make getting around much easier.

The reports indicate that most falls occur when riders fail to use the provided handrail. They can easily slip on the moving stairs due to wet or slippery surfaces, failure to notice debris or worn steps, among other hazards. Falls also occur when riders attempt to carry or handle a heavy or bulky item while riding, causing them to lose their balance and fall. Clearly, the most at-risk groups are the elderly, due to decreased balance, and children, who may not be paying attention. Escalators have also been known to come to an unexpected, sudden stop, and riders not using the handrail can fall without warning.

I include the topic of escalators only to emphasize that this clearly shows how RH can affect individual risk tolerance and acceptance. Our perception of risk decreases because we become comfortable riding the escalator without using the handrails. After all, we've done this many times, and nothing has happened, so it must be okay. Or is it?

How RH Affects the Petrochemical Industry

I mentioned seat belts early in this chapter and a few times throughout this textbook. In the textbook, I compared seat belts to the use of a hydrogen sulfide detector. What I reported there is that when I get into a vehicle, the first thing I do, even before starting the vehicle, is put on my seat belt. This is a learned response, and I'm glad to say that my kids and grandkids do the same. I realize that my chances of being in a serious automobile accident are very low. However, I also acknowledge that if it does happen, the consequences can be catastrophic for me.

The potential exposure to hydrogen sulfide in a refinery and most petrochemical plants can also be catastrophic. It can easily result in the loss of life for many people. It's kind of like driving; it doesn't happen often, and a person could enter the process unit numerous times without consequence. However, when it happens to that person, it could be the last time they enter the unit. I believe this clearly illustrates the hazard of RH.

I briefly reported on the following tragedy in my first Process Safety textbook *Process Operations Safety: The What, Why, and How Behind Safe Petrochemical Plant Operations* and again in more detail in *Hydrogen Sulfide (H_2S) Hazards: Risk Recognition and Prevention*. In this incident, two people lost their lives after being poisoned by hydrogen sulfide gas. Jacob (44) and Natalee were husband and wife, and Jacob received a call from his employer, Aghorn Energy, requesting him to check a pump house.

After Jacob's wife, Natalee, did not hear from him, she took her two young children in the car and drove to the company's pump house to check on him. When she arrived, the gate was unlocked, so she parked near the pump house, left the children in the car, and went inside to see if Jacob was there. As she entered, she was quickly overcome by hydrogen sulfide (H_2S) and collapsed beside Jacob, where the deadly gas also killed her.

An Aghorn supervisor later called law enforcement to check on Jacob since they had not heard from him. Deputies arrived within a few minutes and discovered that Jacob and Natalee had both died from H_2S exposure. The children were found in Natalee's car and were checked at the local hospital, where they were determined to be okay.

What makes this incident so tragic is that it should never have happened. The sour water waste pump inside the pumphouse had failed, releasing high concentrations of H_2S sour water into the building. The responders found the operator's hydrogen sulfide detector on the front seat of his vehicle, and it was in alarm mode. Had Jacob been wearing his detector, it could have saved both lives. It was also reported that the in-situ H_2S detector in the building was not functional.

This is exactly what we are talking about here: the effects of RH. I have no way of knowing if this was the first time the operator had entered this pumphouse without his H_2S detector, but I don't like the answer I get when I ask myself this question. Remember the seat belt analogy above – use it each time and every time. It can save your life. And these rules must be in place and enforced.

Several Types of Routine Activities Where This Has Occurred

There are several types of routine operational tasks where RH has set in. These tasks are usually performed regularly, sometimes daily, and in some cases, multiple times during a single shift. In most of these tasks, written procedures either do not exist or are not consistently followed because the operator is well-trained, experienced, and has performed these tasks many times as part of their routine work.

However, due to RH, they often fail to recognize the risks or hazards. The steps may not be fully completed, may be done in the wrong order, or they may perform the task using incorrect equipment or systems.

The following are just a few examples of these types of tasks that have occurred in the refining and petrochemical industries and continue to happen today.

Examples of RH in Petroleum and Petrochemical Operations or Process Organizations

- Isolating equipment for maintenance involves examples where the equipment was prepared and isolated using the lockout and tagout process. This has occurred multiple times, especially when the pumps are identical and numbered similarly, but each is designated with a suffix, such as the "A" pump or the "B" pump. For instance, two identical pumps in the same service, one labeled P-101A and the other P-101B. P-101B was to be isolated for maintenance; the operator drained, isolated, and locked out P-101A. When the workers arrived, they had a work permit issued to open and conduct maintenance on P-101B, which was still pressurized with process fluids!

- A similar example is when the operator is to lock out an electrical breaker so that electricians can disconnect the power leads to a large motor, such as for a pump, compressor, or fan. All the breakers are in an electrical substation and are essentially identical in appearance and construction. The difference, of course, is that they are labeled with the equipment identification. A similar situation occurred when the operator racked out the wrong breaker, and in some cases, unfortunately, the breaker was in service, causing the equipment to be inadvertently shut down.
- Other examples are where the operator is to switch a crude feed tank to an operating unit or switch a product stream to a different tank and they inadvertently line up the piping circuit from or to the wrong tank. This has resulted in the loss of feed to the crude unit and the overfill of the product tank.

Examples of RH in Mechanical or Maintenance Organizations

- This book gives examples of craftsmen opening the wrong pipeline with tragic results. Usually, lines leak when flanges are opened, causing hydrocarbons to leak or spray out. Fortunately, in many cases, the craftsmen were able to reclose the flange and stop the leak quickly. However, working on the wrong system always risks a tragic outcome, depending on the service and product involved.
- Craftsmen have also started to open, and in some cases, have opened the manway on the wrong tank. As shown in the example above, this creates a significant risk of containment loss, fire, and even explosion. These last two cases can result from RH if the equipment number or other identification is not verified on the work permit or in their work plan documentation.
- Sometimes, craftsmen select the wrong part, such as a bearing or seal, for a pump or other equipment, which can cause premature failure once the equipment is put back into service. RH can also contribute to this because it is a routine and repetitive part of their role, and often, formal procedures might not be in place. Typically, parts should be checked against the equipment records, manufacturers' recommendations, and parts catalogs, among other sources.

The main point is that we are all susceptible to RH when performing tasks or operations repeatedly. I have observed that procedures, when available, can help reduce RH. However, regular training is also essential to maintain the awareness necessary for safe performance, preventing errors that could cause serious consequences. I also believe that periodic reviews of procedures are beneficial and should be conducted.

For operations groups, the US Occupational Safety and Health Administration Process Safety Standard (OSHA) 29 CFR 1910.119 requires a review of operating procedures every three years. I believe that, in some cases, especially on large or complex process units, this review should be conducted more frequently to ensure operational proficiency.

Addressing the Hazard of RH

The studies on RH suggest that repeatedly ignoring safety rules or taking shortcuts occurs over an extended period. However, this does not entirely convince me. I recall the incident at the Pemex Refinery in Deer Park, Texas, on October 10, 2024. This incident is discussed in Chapter 11 of this book, where a major H_2S release happened when workers were opening flanges to install blinds and prepare the pipeline for maintenance. Two workers were killed, and the cities of Deer Park and Pasadena were ordered to shelter in place.

The CSB report is still pending, but it's clear that the workers opened a blind on a different pipeline than the one they were supposed to. The lines ran parallel in the pipe rack, and they opened an in-service line with extremely high levels of hydrogen sulfide (H_2S). My point is that the workers had been working for hours, not days, weeks, or months. As they moved from flange to flange, opening the flanges, they mistakenly accessed the wrong pipeline. Were they not paying attention, or did they become complacent, and did RH play a role?

RH happens when the brain gets used to repeated exposures, especially during daily tasks repeated multiple times without noticeable or felt consequences. When no one gets hurt or injured, the person's perception of risk decreases. Studies describe this as a dangerous survival mechanism that causes workers to take risks they normally wouldn't. The problem is that the threat stays the same, but our reaction to it changes, making RH a concern.

Studies suggest that the first step is recognizing that RH is a normal function occurring in our subconscious. Without awareness, it can happen to anyone. Without this recognition, we might work in high-risk environments and fail to identify hazards or take steps to protect ourselves. The next time you step on an escalator, I hope you recall this brief chapter and use the handrail.

Studies also emphasize the importance of training. Training should address the hazards of RH by explaining that it is an issue and how it develops, to boost workers' awareness. Additionally, studies show that virtual reality with accident simulations can trigger an emotional response to fake accidents, making workers more alert and cautious.

Additionally, studies show that employee safety programs should be designed to help workers understand and stay aware of their personal risk tolerance and how RH is woven into procedures and practices. This can be accomplished through bulletins, safety alerts, periodic reviews of industry incidents where procedures or shortcuts caused accidents and similar measures.

End-of-Chapter Quiz

1 Which of the following best describes "risk perception"?
  Please select the most correct answer.
  - a Risk perception is how we each judge the level of danger from a specific hazard.
  - b Risk perception is how we respond to risks we identify while working.
  - c Risk perception is how our team works together to mitigate or eliminate risks.
  - d Risk perception is the unsafe acts I see others doing while at work.

2 Please describe in your own words what the term "risk habituation" means.
  - a Risk habituation refers to the risks we assume at work when things aren't going well.
  - b Risk habituation refers to the risks we assume when our leadership is absent.
  - c Risk habituation is a term describing how we become accustomed to risks through repeated exposure while working in dangerous environments.
  - d Risk habituation is a process used to help us identify and evaluate risks.

3 Please describe the difference between risk perception and risk habituation.

4 Studies show that risk habituation can be improved through bulletins, _________ alerts, periodic reviews of industry _________ and similar measures where procedures or shortcuts caused ___________.

5 Studies also emphasize the importance of __________. Training should address the hazards of risk habituation by explaining that it is an issue and how it __________, to boost workers' ___________.

6 For operations groups, the US Occupational Safety and Health Administration Process Safety Standard (OSHA) 29 CFR 1910.119 requires a review of operating procedures every __________________.

7 Please list the three industries or areas mentioned in this Chapter where risk habituation has occurred with frequency.

8 Risk habituation happens when the ________ gets used to repeated __________ to risks, especially during daily tasks repeated multiple times without noticeable or felt consequences. When no one gets ________ or __________, the person's perception of risk decreases.

9 Risk habituation has been described as a dangerous survival mechanism that causes workers to take risks they normally wouldn't take. The problem is that the threat stays the same, but our reaction to it changes, making risk habituation a concern.

Please select true or false to this statement.
True
False

Additional References

ASCE, Reducing Risk Habituation to Struck-By Hazards in a Road Construction Environment Using Virtual Reality Behavioral Intervention, https://par.nsf.gov/servlets/purl/10524452#:~:text=Risk%20habituation%E2%80%94lowered%20alertness%20to,Ahn%202020;%20Oken%20et%20al.

Centers for Disease Control and Prevention (CDC), NIOSH Science Blog, Psychosocial Hazards Often Overlooked in Construction Industry, https://blogs.cdc.gov/niosh-science-blog/2023/02/15/construction-psychosocial-hazards/.

Hadil Haj Ali, Moshe Glickman, and Tali Sharot, The Role of Habituation in Risk-taking Escalation, https://www.2023.ccneuro.org/proceedings/0001245.pdf?s=W&pn=1633.

The National Institute of Health (NIH), National Library of Medicine, National Center for Biotechnology Information, Escalating Risks: Injury Patterns in Escalator-Related Trauma, https://pmc.ncbi.nlm.nih.gov/articles/PMC11671416/.

The National Institute of Health (NIH), National Library of Medicine, National Center for Biotechnology Information, Risk Perception: It's Personal, https://pmc.ncbi.nlm.nih.gov/articles/PMC4181910/.

The National Institute of Health (NIH), National Library of Medicine, National Center for Biotechnology Information, Virtual Accident Curb Risk Habituation in Workers by Restoring Sensory Responses to Real-world Warning, https://pmc.ncbi.nlm.nih.gov/articles/PMC9830218/#:~:text=In%20high%2Drisk%20work%20environments,fatalities%20and%20injuries%20at%20workplaces.

Transportation Consortium of South-Central States, Lead University: Texas A&M University, Preventing Struck-by Hazards: Defying Risk habituation via Virtual Accident Simulation, https://repository.lsu.edu/cgi/viewcontent.cgi?article=1122&context=transet_pubs.

10

Preventing "Normalization of Deviation"

I believe that I understood the concepts about the normalization of deviation, sometimes also referred to as "normalization of deviance," before attending a seminar presented by Mike Mullane titled "Normalization of Deviance." However, he defined normalized deviation more clearly than I had ever heard. While it has been a long time since I attended Mike's seminar, it made an impression and left me with learnings I have never forgotten. Mike Mullane is an engineer and weapon systems officer, a retired USAF officer, and a former NASA astronaut. Mike flew as a mission specialist on several space shuttle missions during his career. He is also an author and a professional speaker.

Simply stated, normalization of deviation is a person, such as a crew member or possibly an entire team, repeatedly deviating from a standardized procedure or practice and failing to recognize the hazards or potential consequences of deviation. This typically occurs when there are schedule demands and time is limited; however, in addition to schedule demands, it can also happen due to other drivers, such as budget compliance, a perceived efficiency improvement, or other perceived benefits without recognition of the potential consequences of the deviation.

Sociologist Dr. Diane Vaughan, another author, also wrote a textbook titled *The Challenger Launch Decision: Risky Technology, Culture, and Deviance at NASA*, where she describes this phenomenon in more detail. Diane is quoted as saying, *"Social normalization of deviance means that people within the organization become so much accustomed to a deviation that they don't consider it as deviant, despite the fact that they far exceed their own rules for the elementary safety."* She was also quoted this way: *"Normalized Deviation is the gradual process through which unacceptable practice or standards become acceptable. As the deviant behavior is repeated without catastrophic results, it becomes the social norm for the organization."*

Sociologist Dr. Diane Vaughan, 1996 (The Challenger Launch Decision, 1996). Please see the book references in the Additional References section.

Note to Readers: a brief discussion of the two shuttle incidents is provided here; more details are provided later in this chapter.

The Rogers Commission investigated the Space Shuttle Challenger explosion. They determined the cause was a design flaw with the solid rocket booster (SRB) O-ring seals, which had also failed several times before the fateful day. However, no damage was sustained during the prior failures, and the failures were apparently not considered serious failures. The shuttle continued to fly even though the prior failures had been documented, with no action taken to improve the seal design or their resistance to cold temperatures.

What was different on the morning of the space shuttle incident was the ambient temperature at the time of launch. Icing had been reported on the launch pad overnight, and the ambient temperature was 36 °F at 11.38 am, the time of launch. It is noteworthy that Morton Thiokol's engineers had previously recommended "not to launch below 53 °F." The Rogers Commission determined that the low ambient temperature contributed to excess leakage of the O-ring seals, causing the fire that impinged on the external fuel tank and the resulting explosion.

Unfortunately, another shuttle and the entire crew were lost again on Saturday, February 1, 2003, when the Space Shuttle Columbia disintegrated as it reentered the atmosphere over Texas and Louisiana. Seven experienced astronauts were on board, and all were lost. Like the Space Shuttle Challenger, this incident was also the result of the normalization of deviation. The shuttle fleet had developed an impressive history of accomplishments at this point, having completed 112 successful missions. This mission was the 113th mission and the 88th mission since the Challenger disaster. During previous launches, foam insulation material had separated from the external fuel tank, causing mostly minor damage, although in a couple of cases, near catastrophic damage to the insulating tiles on the orbiter. In this launch, foam insulation material broke away from the Space Shuttle's external tank and struck the thermal protection system tiles on the orbiter's left wing. This damaged the thermal protection tiles and allowed the super-hot air to enter the wing beneath the thermal insulation, melting the aluminum structures and causing the Space Shuttle to fail.

We should all be aware that deviations can be normalized in any operation, especially those operating critical facilities and/or critical chemicals, such as a petroleum refinery, a petrochemical plant, a nuclear power generation facility, or other similar operations. Anytime deviations are recognized, we must be aware of and insist on a full review and risk assessment before deviating from established norms. This risk assessment process should conform with practices similar to conducting a process hazard analysis (HAZOP), where changes in weather conditions are considered, as well as changes or deviations in process operating conditions such as operating temperature, pressure, and flow rates. Other operating conditions to consider are changes in pH, differential pressure, feedstock, feed properties, etc. Once the approval is obtained to operate with the deviation, this should be formalized with the Management of Change documentation.

Normalization of Deviance applies to many industries, professions and organizations, including medicine, aviation, petroleum refining and petrochemicals, manufacturing, and many others. It has also been characterized as a silent killer because deviations can quickly become the "new way" that things get done without the benefit of risk assessment and consideration of the potential consequences and the mitigation of those consequences. The deviation may bypass or ignore safety devices, systems, or well-established mitigation of an otherwise unforeseen hazard.

The risk is that deviations from safety systems, standards, procedures, or established guidelines are quickly accepted because there are no obvious negative consequences, and no one calls attention to the deviation or the bypassed system or process. However, at some point, there may be a slight difference in the scenario where something as simple as a different weather situation, another piece of equipment already out of service, or something similar can lead to a catastrophic outcome.

It is difficult for an organization to recognize that deviation occurs because it is a slow and gradual change in how things get done. It may be what is perceived to be a minor change where the benefit is quickly recognized, for example, allowing a task to be completed much quicker than following the full procedure. For example, taking a shortcut in a procedure rather than following each step or doing the steps in a different sequence than the written procedure. While taking the shortcut, the job was completed sooner, and there were no observed consequences. No one said, "Wait a minute, let's talk about this since we are not following the procedure," or asked, "What can go wrong if we do it in this new way?" Then, the next time they execute the procedure, the "new way" is done repeatedly without consequence. This reassures the individual and possibly the entire team that nothing bad happened, and the change is slowly accepted and becomes the new unwritten way the task is done. Although the documented procedure is not being followed, no one says anything, and there are no negative outcomes, so it must be OK. This can quickly be perceived as the "new way" even though it has not been reviewed, assessed, or documented, and since there was no opposition, the perception is that everyone on the team has accepted it.

Of course, the real issue with normalized deviance is that we may get away with the "new way" repeatedly until one day, possibly the next time this is done, something may be different that was not considered when the deviation was implemented. The change may have been considered in the original procedure or standard, and mitigation was designed to address this circumstance. It was built into the original standard but has not been considered in the deviation. This can have catastrophic consequences, including loss of life.

Normalization of deviation can occur in any organization but is more likely to occur with a continual focus on improved efficiency. The dictionary defines efficiency as "the quality of being able to do a task successfully, without wasting time or energy." Most people think efficiency is saving time or completing a task more quickly. For example, the potential to miss a scheduled product delivery or the time required to do equipment checks fully can lead to people making shortcuts in procedures or standards or deferring scheduled equipment checks. Unfortunately, these decisions may be made for the right reasons. However, failure to fully consider the potential consequences may lead to catastrophic results. The concern is that this can lead to people taking shortcuts without fully considering the potential consequences of their actions. This leads to the "new way" being unconsciously accepted by the team even though it has not been reviewed and documented.

The following concerning the BP Texas City incident and the two NASA case studies were included in my first Wiley textbook titled *Process Operations Safety: The What, Why, and How Behind Safe Petrochemical Plant Operations*. They are included here due to the significance of the events and the key lessons learned from each.

Did Normalized Deviation Play a Role at BP Texas City?

The refinery explosion occurred on March 23, 2005, killing 15 workers and injuring 180 others. The explosion occurred during a unit startup when a process fractionation column was overfilled, releasing light naphtha into a process area where multiple ignition sources existed. The resulting vapor release resulted in significant vapor cloud explosions, resulting in fatalities, injuries, and catastrophic damage to the unit and surrounding infrastructure.

Based on the investigation completed by the US Chemical Safety and Hazard Investigation Board (CSB), it appears that normalized deviation played a significant role in the BP Texas City explosion and fire of March 23, 2005. BP Texas City had written procedures for establishing tower levels during startup operations; however, the CSB found that operators routinely deviated from these procedures. The CSB investigation reported, "In the majority of the 19 ISOM startups between April 2000 and March 2005, the tower was filled above the range of the level transmitter despite procedural instructions to fill the tower to a 50 percent reading on the transmitter. Operators frequently ran the valve sending liquid raffinate out of the unit to storage in "manual" instead of the "automatic" control mode as required by the procedure."

This is only one example of procedural deviations reported by the CSB in their report. From the report, it is easy to conclude that the operators had deviated from the procedure so many times that the deviation had become "normal," or the way things are done, rather than following the procedures. This deviation is a dangerous practice because when things are not what is expected, or things are different on that day or at that time, the results can be catastrophic. At BP, the routine practice of running above the range of the level instrument on March 23, 2005, coincided with several other instrumentation and alarm failures and resulted in the overfill of the tower and the resulting catastrophic explosion and fire.

Unfortunately, other examples from the BP Texas City incident also exemplify normalized deviation. For example, the operator on the ISOM unit left the post an hour before the end of the shift without making a turnover with their replacement, and turnover at the console position was marginal at best. During their investigation, the CSB found that there was "widespread tolerance of non-compliance with basic HSE rules"; in other words, at BP Texas City, not following the rules was accepted ("tolerated"). I believe this speaks volumes about the organizational culture (more on this later).

Frequently, operators may deviate from written procedures with the best intentions, thinking they are taking the safest path. However, they may not have the benefit of knowing the reason behind the written procedure, or they may not be fully aware of other conditions in the unit. Strict adherence to written procedures is the best course of action. If a concern exists, they should always discuss it with their supervisor and jointly decide on the best course of action. If the procedure is revised, it should be "redlined" with the changes, and a supervisor should review and approve the changes before proceeding.

Lessons from NASA Shuttles Challenger and Columbia Disasters

The Challenger Space Shuttle Disaster

Many of us will remember, or we have certainly heard the story, that on January 28, 1986, the Shuttle Challenger exploded 73 seconds into its launch, killing all 7 crew members. The subsequent NASA investigation determined that a SRB O-ring seal in one of the rocket booster joints failed, resulting in a fire that allowed flames to impinge on the external fuel tank. The impinging fire resulted in the rupture of a liquid hydrogen tank, which exploded, causing the rupture of a liquid oxygen tank. The resulting massive explosion destroyed the shuttle and resulted in the loss of the entire crew.

Before the Challenger disaster, NASA had flown 24 shuttle missions with 4 different vehicles without incident. However, the investigation revealed that the O-rings had malfunctioned (leaked) on seven missions, although without incident. In hindsight, it was evident that the NASA team had expected that the O-rings may leak and accepted this as "normal." NASA knew that the Morton Thiokol design of the O-ring seal was flawed nearly nine years before the loss of the Challenger but had not addressed the concern. On the morning of the tragedy, Florida's ambient temperature was frigid (31 °F or −6 °C), and the engineers cautioned about continuing with the mission. NASA management did not accept these warnings, and the mission deviations were accepted as normal, and the shuttle was launched. In hindsight, the investigation revealed that the O-ring seals leaked on all previous launches where the ambient temperature was below 65 °F (18 °C). See Figure 10.1 for images of the weather impacts on the morning of the Challenger launch.

Normalized deviation played a role in the Challenger incident. Had the flaws with the O-ring seals been addressed earlier, or had the mission been delayed due to the cold weather, the Challenger incident may not have occurred, and the seven astronauts' lives would not have been lost.

- Launch day temperatures as low as 22 °F at Kennedy Space Center.
- Thiokol engineers had concerns about launching due to the effect of low temperature on O-rings.
- NASA Program personnel pressured Thiokol to agree to the launch.

NASA/Public Domain.

Figure 10.1 Image of the weather impacts on the morning of the Challenger Launch. The Challenger exploded after 73 seconds of flight. All seven astronauts were killed when the cabin crashed into the ocean.

The following "contributing factors" were part of a report on the Challenger disaster published by NASA:

> **Normalization of Deviance:** The space shuttle's SRB problem began with the faulty design of its joint and increased as both NASA and contractor management first failed to recognize it as a problem, then failed to fix it, and finally treated it as an acceptable flight risk*.
>
> **Organizational Silence:** The decision to launch Challenger was flawed. Those who made that decision were unaware of the recent history of problems concerning the O-rings and the joint and were unaware of the initial written recommendation of the contractor advising against the launch at temperatures below 53 °F and the continuing opposition of the engineers at Thiokol after management reversed its position.
>
> **Silent Safety Organization:** There were serious ongoing weaknesses in the shuttle Safety, Reliability, and Quality Assurance Program, which had failed to exercise control over the problem tracking systems, had not critiqued the engineering analysis advanced as an explanation of the SRM seal problem, and did not provide the independent perspective required by senior NASA managers at Flight Readiness Reviews.
>
> **Boston College sociology professor Diane Vaughan, author of the book "The Challenger Launch Decision," referred to this as "The normalization of the technical deviation of the booster joints ..."*

The Columbia Space Shuttle Disaster

Seventeen years after the loss of the Challenger, a similar series of circumstances surfaced in another NASA incident. On February 1, 2003, the space shuttle Columbia disintegrated while reentering Earth's atmosphere at 10,000 mph, scattering debris over 2,000 sq. miles of Texas, again with the loss of all 7 crew members and destruction of the $4 billion spacecraft. As a result of this incident, NASA grounded the entire shuttle fleet for 2½ years for investigation and modifications to the fleet.

This time, the investigation revealed that during the launch of shuttle Columbia on its 28th mission to space, a piece of foam insulation detached from the external tank, striking the shuttle's left wing. A similar loss of foam insulation had also occurred during previous shuttle launches, some with nearly catastrophic consequences. However, even with the prior incidents, the shuttle was allowed to continue to fly. Upon reentry to Earth's atmosphere, the failed insulation allowed the super-hot atmospheric gases to penetrate the heat shield, destroying the wing structure and disintegrating the shuttle.

Again, normalized deviation appears to play a role in the Challenger incident. Had the issue with the foam insulation been addressed before this mission, or had

NASA/Public Domain.

Figure 10.2 The Challenger crew (In memorial). STS-51L Crew (l-r): Payload Specialists Christa McAuliffe and Gregory B. Jarvis, Mission Specialist Judith A. Resnik, Commander Francis R. Scobee, Mission Specialist Ronald E. McNair, Pilot Michael J. Smith, and Mission Specialist Ellison S. Onizuka.

management postponed the mission until it was addressed, the Columbia incident may not have occurred. The Columbia Accident Investigation Board reported, "Cultural traits and organizational practices detrimental to safety were allowed to develop," "reliance on past success as a substitute for sound engineering practices," and "organizational barriers that prevented effective communication of critical safety information." I believe these are all symptoms of normalized deviation.

Please refer to Figure 10.2 for a memorial image of the Challenger crew and to Figure 10.3 for a memorial image of the Columbia crew. These men and women were truly pioneers in space, advancing our technology for the future. We must all learn from these tragic incidents and how they can be applied to our industry.

Key Lessons Learned

The following is from the same NASA report on the Challenger disaster:

- We cannot become complacent.
- We cannot be silent when we see something we feel is unsafe.
- We must allow people to come forward with their concerns without fear of repercussion.

Figure 10.3 The Columbia crew (In memorial). The STS-107 crew. Seated in front are astronauts Rick Husband, commander, and Willie McCool, pilot. Standing (from left) mission specialists Dave Brown, Laurel Clark, Kalpana Chawla, Mike Anderson (payload commander), and payload specialist Ilan Ramon, representing the Israeli Space Agency.

How to Prevent the Normalization of Deviance?

It should be accepted that deviation will be challenged. When driving on the interstate, I typically drive at about the same speed as the traffic. In most states, the interstate speed limit of 70 mph has been adopted. However, the traffic normally flows outside the city in open areas at around 75–80 mph. So, what is the new accepted speed limit? Is it 70 mph, or is it 75–80 mph? If I get stopped by a state police officer for going 85 mph, am I going 5 mph above the speed limit or 15 mph above the limit? This is an example of the normalization of deviance. In this hypothetical scenario, I accepted the new speed limit to be 80 mph and felt I was only going 5 mph above this limit. How do you think the state police officer would view this argument? I should have expected to receive that ticket for speeding by driving 85 mph in a 70 mph zone.

Normalizing deviance can lead us all down a path where bad things can happen. For example, accepting a reactor temperature that is very slowly creeping upward could easily be accepted or possibly not even noticed over a longer period.

Of course, this can result in a violent temperature runaway in a chemical reaction and an explosion that severely damages or destroys equipment and injures many people.

Ensure your organization and team have a strong safety culture. This culture should support and encourage reporting deviations and emphasize the importance of following standard operations procedures or practices, even when there are pressures to deviate. These should be noticed, and when an employee identifies or brings up a deviation that may have the potential for serious consequences, the employee should be recognized or rewarded. This will have the effect of communicating that this is an expectation for all employees to follow suit.

One way I have seen this addressed is to designate one of the key people in the team to be a team champion for the initiative or system. For example, we implemented a new alarm management initiative at one of our sites. It just so happened that I was selected to champion this process and guide it to completion. This complex process involved many others, including disciplines from operations, technical, and computing. We worked together as a team and completed the project well within the expected timeline, and the results speak for themselves. This illustrated the importance of a team champion, not necessarily a leader, but someone to guide the process.

Managers have a strong role to play in this regard. For example, when setting operations targets, the targets must be realistic and allow all team members to have input into the process. Otherwise, if targets are unrealistic, this can impose production pressures, resulting in shortcuts. Of course, shortcuts can unexpectedly impact the operation.

Likewise, if you or another team member notice another employee accepting a deviation as "the new normal," this should also be brought to the employee's attention, along with the process or procedure for how this can become the new normal. For example, discuss how the new procedure must go through the management of change process and how potential risks are fully evaluated and properly mitigated.

Once recognized, deviation should never be allowed to continue – it must be addressed or formally reviewed and accepted. This is important since deviation will continue until conditions are different and an unexpected outcome can occur. Also, deviation occurring in one section of the facility may spread to a different section where conditions may differ.

Communication between team members must always be clear, concise, and focused on the topic being discussed. All team members should feel free and encouraged to bring up issues and concerns, which should be discussed with input from all participants. Ensure all employees know the consequences of non-compliance and feel empowered to speak up if they see unsafe practices or are aware of deviations from company policies or procedures. Continuously monitor adherence to standard operating practices (SOPs) and provide feedback to teams on identified deviations or issues before they become normalized.

There is a link to a very interesting article by Mark Paradies of the TapRoot® organization posted in the Additional References section of this Chapter titled "Stopping the Normalization of Deviation with the Normalization of Excellence – How Admiral Rickover Did It." I encourage readers of this book to read this and consider the TapRoot training course mentioned.

End-of-Chapter Quiz

1 Please describe what "normalized deviation" means in your own words.

2 At BP Texas City, the operators were reportedly deviating from the operation procedure routinely. Understanding the description of normalized deviation, could this have been occurring at that site, and could it have contributed to this catastrophic process safety incident?

3 How is normalized deviation reinforced?

4 You are aware that on January 28, 1986, the Shuttle Challenger exploded 73 seconds into its launch, killing all 7 crew members.
What role did normalized deviation plan in this tragic accident?

5 Seventeen years after the loss of the Challenger, on February 1, 2003, the $4 billion Columbia was also lost, and the entire crew was killed. This incident resulted in NASA grounding the entire shuttle fleet for 2½ years.
Do you remember that normalized deviation played a role in this tragedy?

6 How can normalized deviation be recognized and challenged in your organization?
Please provide examples of how this can be done.

Additional References

FAA Aviation Safety - FAST Team – Safer Skys Through Education, CFIT and Normalization of Deviation Bias, https://medium.com/faa/cfit-and-normalization-of-deviation-bias-5f6e8da61c4f.

Flight Safety Foundation, Chris Sharber, a first officer and flight simulator instructor–Boeing 777 fleet, at the United Airlines Training Center, Normalization of Deviance, https://flightsafety.org/asw-article/normalization-of-deviance/.

IChemE, Loss Prevention Bulletin, Improving Process Safety by Sharing Experience, April 2020, https://www.icheme.org/media/13597/lpb272online.pdf.

Industrial Safety and Hygiene News (ISHN), Reversing the Normalization of Performance Deviations Can Be Difficult, https://www.ishn.com/articles/114423-reversing-the-normalization-of-performance-deviations-can-be-difficult.

Inspectioneering, Posted by Inspector Frank on October 5, 2021, Let's Be Frank: Stop Normalizing Deviance, https://inspectioneering.com/blog/2021-08-26/9796/lets-be-frank--stop-normalizing-deviance.

National Aeronautics and Space Administration, The Cost of Silence: Normalization of Deviance and Groupthink (Senior Management ViTS Meeting) November 3, 2014, https://sma.nasa.gov/docs/default-source/safety-messages/safetymessage-normalizationofdeviance-2014-11-03b.pdf.

National Library of Medicine, When Doing Wrong Feels So Right: Normalization of Deviance, https://pubmed.ncbi.nlm.nih.gov/25742063/.

Mark Paradies (April 29, 2024), Stopping the Normalization of Deviation with the Normalization of Excellence – How Admiral Rickover Did It, https://taproot.com/stopping-the-normalization-of-deviation-with-the-normalization-of-excellence-how-admiral-rickover-did-it/.

Psych Safety, Normalisation of Deviance, https://psychsafety.com/normalisation-of-deviance/.

US Chemical Safety and Hazard Investigation Board (CSB), Key Lessons for Preventing Incidents When Preparing Process Equipment for Maintenance, https://www.csb.gov/delaware-city-refining-company/.

Diane Vaughan, The Challenger Launch Decision: Risky Technology, Culture, and Deviance at NASA, https://books.google.com/books?id=erYjCwAAQBAJ&pg=PT30#v=onepage&q&f=false.

Vervint, Normalization of Deviance: Definition, Examples and Solutions, https://vervint.com/article/normalization-of-deviance-definition-examples-and-solutions/.

Youtube, Mike Mullane | Nominalization of Deviance | Keynote Speaker | SpeakInc, https://www.youtube.com/watch?v=GN80sx3s4LA.

11

Opening Process Equipment: What Has Happened/ What Can Happen and How Is It Prevented?

Introduction

Chapter 11 addresses incidents within the petroleum refining or petrochemical industries that involve or are related to opening process equipment (OPE). Essentially, all these incidents also pertain to the loss of containment of flammable or toxic materials and can have devastating effects on personnel conducting the work, others at the facility, or even individuals in surrounding communities.

Like most of this textbook, this chapter is based on reports from US federal agencies and agencies from other governments, including the US Occupational Safety and Health Administration (OSHA), the US Chemical Safety and Hazard Investigation Board (CSB), the British Health and Safety Executive (HSE), various US state agencies, and additional sources like legal proceedings and publicly available media reports.

In several of these reports, I have also included the Author's Comments, which reflect my viewpoint on the incident. These comments are grounded in my over 50 years of experience in the US refining industry and my training and expertise in personnel and process safety. They express my perspective and are not meant to imply additional meanings or suggest that these comments carry legal obligations.

The primary goal here is to share the lessons learned from these incidents to prevent similar occurrences elsewhere and to minimize the risk of others becoming severely injured or worse. I believe these books will be utilized in community colleges, trade schools, and refining and petrochemical facilities to help train their employees in process safety and lessons learned from process safety incidents. There is no other objective, and I am not interested in placing blame or harboring resentment toward the companies or facilities involved.

At the end of the discussion of each incident, there is a section called "Additional References." This section provides links to incident reports or related resources from other agencies, providing additional access to more detailed information about each incident.

The Chapter 11 incident discussion is a series of Case Studies, organized by incident cause (thematic clarity) and the date of occurrence.

Series I. Loss of Containment Due to the Failure to Identify Hazards (Hazard Identification)

Series II. Loss of Containment Due to the Failure to Properly Identify Equipment for Opening or the Workers Cutting or Opening the Wrong Equipment

Series III. Loss of Containment Due to the Failure to Follow Procedures or Noncompliance with Work Permit Requirements

Series IV. Loss of Containment Due to the Failure to Achieve Energy Isolation

Series V. Explosion, Fire, or Loss of Life Due to Inadequate or No Gas Testing

There are two very recent incidents and one somewhat recent incident, all involving OPE. Chapter 11 is titled "What Has Happened/What Can Happen," and I strongly believe that unless we learn from past incidents or have precautions in place, if it has happened before, it can happen again.

The first incident happened on October 24, 2024, at the PERMEX Refinery in Deer Park, Texas. Mechanics were opening flanges on a long pipeline to install isolation blinds for turnaround activities when they accidentally opened a flange on the wrong pipeline. This caused a large release of highly concentrated hydrogen sulfide gas. One worker involved in the line opening died, as did another worker who was working downwind from the flange opening. Thirteen others sought medical attention due to H_2S exposure, and the nearby cities of Deer Park and Pasadena were ordered to shelter in place for several hours after the release.

The next case study covers an incident at the Imperial Oil Refinery in Sarnia, Canada, on April 2, 2019. Unfortunately, similar incidents have occurred at refining and chemical plant sites across the United States and other regions. When a hydrotreater fractionation column was opened for maintenance before a planned turnaround, a fire broke out inside the column because of the buildup of pyrophoric materials. During the fire, the tower collapsed and struck another column as it fell. Fortunately, no injuries occurred, although the column was destroyed, resulting in extended downtime for its replacement. Many lessons can be learned from this incident, which will be discussed in this case study.

The third of these introductory case studies occurred at the PBF Martinez Refinery in California on Saturday, February 1, 2025, when a hydrocarbon release and a major fire happened while workers were opening equipment for maintenance. According to local media, the fire caused six minor injuries, four of which were treated at a nearby hospital before being released. The large fire released significant smoke plumes into the community, leading to a shelter-in-place order and a public health advisory. Contra Costa Health Services is leading the investigation and has hired an engineering and technical firm, which reported that the workers accidentally opened the wrong flange, causing the hydrocarbon release and fire. Local media also reported that the Martinez Refinery has shut down due to damage from the large fire.

These three case studies are all included in Chapter 11, where you'll find additional similar incidents with many lessons to learn.

Series I. Loss of Containment Due to the Failure to Identify Hazards (Hazard Identification)

Incident 1: Loss of Containment Due to the Failure to Identify Hazards (Hazard Identification)

BP-Husky, Oregon Refinery, Toledo, Ohio

On September 20, 2022, two brothers were working as outside operators at the BP-Husky Toledo Refinery when the refinery was undergoing an upset condition. The Crude Unit 1 fuel gas mix drum is filled with light naphtha. The investigation revealed that the two outside operators removed a 2-inch blind and opened valves to drain the naphtha from the fuel mix drum directly to the ground. Shortly afterward, the wind direction changed, and the Crude Unit 1 furnace ignited the vapor cloud.

September 20, 2022; two fatalities (two brothers were killed in this tragic incident)

This incident was also investigated by the CSB, and the following provides insight into the CSB report. The following contains excerpts from the CSB final report on this incident:

> The Naphtha release and fire at the BP-Husky Toledo Refinery on September 20, 2022, tragically resulted in the deaths of two employees, brothers Ben Morrissey and Max Morrissey.

At the time of this incident, BP operated the refinery in Oregon, Ohio, east of Toledo. It is now owned and operated by Ohio Refining Company, LLC (ORC), a subsidiary of Cenovus Energy Inc. The CSB report and this summary refer to the refinery as the BP Toledo Refinery. This summary has been edited to save space and includes excerpts from the CSB final report.

Figure 11.1 illustrates the large refinery fire that occurred due to the release of the flammable materials when the operator drained the fuel gas mixing drum to grade.

Refinery Events Leading to This Incident (Taken from the CSB Final Report)

The CSB found that on the evening of September 20, 2022, when the incoming shift arrived, the refinery was experiencing disruptions caused by a series of events that began roughly 24 hours earlier. It all started with a minor process upset the previous evening, which led to additional and increasingly severe upsets. The following briefly describes the refinery upset that occurred and the high level of naphtha that accumulated in the fuel gas mixing drum.

US Chemical Safety and Hazard Investigation Board (CSB). Public Domain.

Figure 11.1 The BP-Husky Refinery, September 20, 2022.

How the Fuel Gas Mixing Drum, Which Is Normally Fuel Gas, Accumulated Significant Amounts of Light Naphtha

According to the CSB, water started accumulating in the Crude 1 Overhead Accumulator Drum after 7:00 p.m. during the night shift on September 19, 2022. Several hours later, it overflowed into the naphtha stream that usually exits the drum. Excess water in the naphtha stream then built up downstream in the Coker Gas Plant Foul Condensate Draw-Off Drum, which eventually overflowed into the Coker Gas Plant Absorber Stripper Tower. This overflow, along with the liquid flow from the Absorber Stripper Tower bottoms into downstream equipment, led to an increase in pressure downstream in the Naphtha Hydrotreater (NHT) preheat system.

During the day shift on Tuesday, September 20, 2022, an increase in pressure within the NHT Preheat caused two emergency pressure-relief valves to open shortly after 7:00 a.m. This led to severe vibrations in the piping, which caused a failure of a ¾-inch drain line connected to the main naphtha piping. The piping failure resulted in a loss of containment of liquid naphtha. Although the naphtha did not ignite, it triggered an emergency shutdown of the NHT unit and a bypass of the Coker Gas Plant. The Crude 1 unit continued operating.

The NHT unit's emergency shutdown caused a Crude 1 Tower upset for much of the Tuesday day shift. With the NHT unit offline and the Coker Gas Plant bypassed, naphtha from the Crude 1 Overhead Accumulator Drum could only be sent to Light Virgin Naphtha Storage. Following the NHT shutdown, the Crude 1 Tower experienced multiple issues, including several losses of pumparound cooling, a level upset in the Crude 1 Overhead Accumulator Drum, and 11 instances of Crude 1 Tower overpressure during the day shift.

As operations personnel worked throughout the day to stabilize the Crude 1 Tower, they made several adjustments to the tower's feed. At 4:56 p.m., the oncoming Tuesday night shift made another change, removing all light crude oil from the Crude 1 Tower feed.

During Tuesday night shift, another Crude 1 Tower process upset occurred due to the rapid and complete loss of light crude oil feed. The upset caused a high liquid level in the Crude 1 Overhead Accumulator Drum. To control the quickly rising liquid level, board operators began transferring the excess liquid to the Coker Gas Plant Absorber Stripper Tower.

Once the board operator intentionally opened the flow control valve from the Crude 1 Overhead Accumulator Drum to the Coker Gas Plant (the "naphtha flow control valve to the Coker Gas Plant"), liquid naphtha flowed from the drum to the Absorber Stripper Tower. With the naphtha flow control valve to the Coker Gas Plant open, Naphtha started filling the Coker Gas Plant Absorber Stripper Tower. Eventually, it overflowed through the Coker Gas Plant bypass line into the Fuel Gas Mix Drum. Once the Fuel Gas Mix Drum was filled with liquid, naphtha flowed to the downstream furnaces and boilers, which are designed to burn only fuel gas. A simplified drawing of the Fuel Gas Mixing Drum is shown in Figure 11.2.

Figure 11.3 illustrates the effects of a cascade of liquid buildup and overflow. This diagram, from the final CSB report on the incident, shows the overflow of liquid naphtha from the Crude 1 overhead accumulator drum to the absorber stripper tower, the overflow from the absorber stripper tower to the fuel gas mix drum, and the flow to the furnaces and boilers downstream of the fuel gas mix drum.

The Incident

At approximately 6:09 p.m., the console operator observed that the level of the fuel gas mix drum was rising and saw a level alarm on the distributed control system (DCS) alarm screen. A few minutes later, the console operator radioed the outside operators to check the fuel gas mix drum level. Four outside operations personnel responded and arrived at the fuel gas mix drum. Another outside operator began draining the fuel gas mix drum into the flare knockout drum. This information indicates that the fuel gas mix drum was filling with light naphtha.

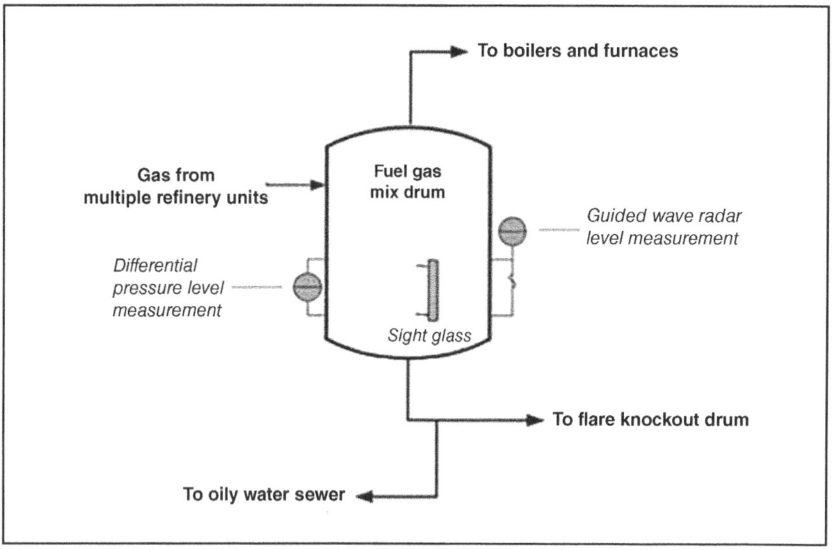

Figure 11.2 A simplified drawing of the fuel gas mixing drum at the BP-Husky Refinery. US Chemical Safety and Hazard Investigation Board (CSB). Public Domain.

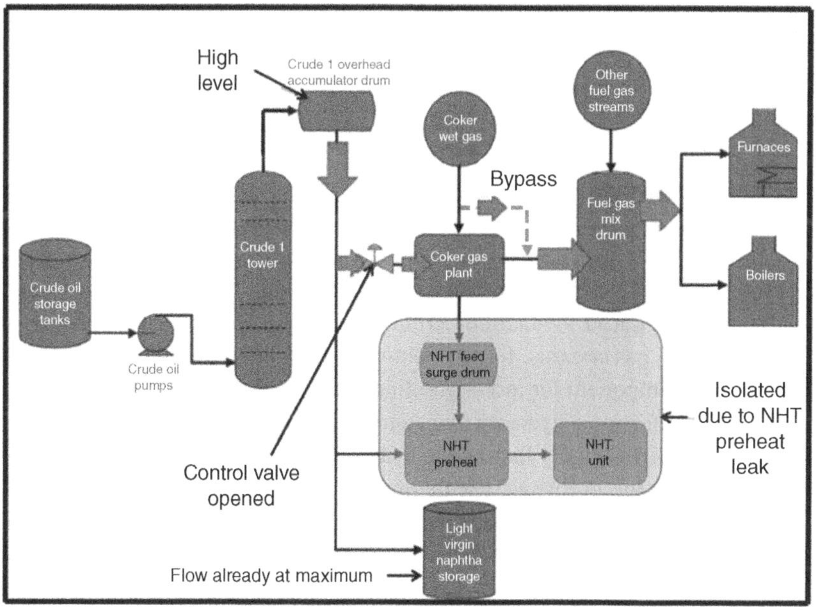

Figure 11.3 Unit sketch by CSB highlighting the cascading liquid accumulation and overflows leading to the tragic loss of life incident. US Chemical Safety and Hazard Investigation Board (CSB). Public Domain.

The outside operators tried to empty the fuel gas mix drum by following these procedures:

- Four operators were present when someone fully opened a 2-inch valve to drain liquid from the fuel mix drum to the flare knockout drum, a closed system.
- One of the operators also opened a 2-inch drain from the fuel mix drum to the oily water sewer, which is a system designed for this purpose.
- Later, a second 1-inch drain was opened from the fuel mix drum guided wave radar level transmitter to the Oily Water Sewer.
- Two of the operations personnel left the area to check on the sweet gas knock out pot in the coker gas plant. This left the two BP employees, the two brothers, who were wearing self-contained breathing apparatus (SCBA) and a hydrogen sulfide (H_2S) detector, to finish draining the fuel gas mix drum. They opened two additional ¾-inch bleed valves to the ground.
- Although no one else was aware, the two brothers began draining liquid light naphtha from the fuel gas mix drum directly onto the ground through a 2-inch valve on the side of the drum. These two BP employees removed the blind flange from the 2-inch valve to release the liquid from the fuel gas mix drum.

Unfortunately, and as I believe was entirely predictable, the naphtha flashed, trapping the two operators (the two brothers) in a large naphtha fire – a fire very similar to gasoline. The CSB reported that the naphtha vapors most likely reached the Crude 1 unit furnace and quickly flashed back to the source, the area surrounding the fuel gas mix drum.

What Is Light Naphtha?

To fully understand the impact of the decisions made on this day and the importance of this incident, one must understand the properties of light naphtha. Light naphtha is a mixture of hydrocarbons containing about five to six carbon atoms. It is typically produced by distilling crude oil, condensates from natural gas, or other petroleum gas streams. In petroleum refining, light naphtha is often used as a blending component for motor gasoline and as a feedstock for other refining units, such as reforming, hydrotreating, and hydrocracking.

Naphtha is a light, highly flammable hydrocarbon mixture mainly made up of pentane and hexane derivatives, with smaller amounts of heavier compounds. Light naphtha typically has a flash point well below 0 °F, similar to gasoline, and a boiling point range between 86 °F (30 °C) and 190 °F (90 °C).

Light naphtha in the fuel gas mix drum also contains significant amounts of hydrogen sulfide (H_2S), which is highly toxic and can be lethal at relatively low concentrations. At low levels, H_2S is easily detected by its rotten egg odor.

However, an H_2S level at or above 100 ppm is considered by OSHA to be Immediately Dangerous to Life and Health (IDLH). About 100–150 ppm can deaden the olfactory nerves, causing the sense of smell to be lost. At this level, a person might assume the gas has dissipated and that it is safe to continue working. See the Wiley book *The Hazards of Hydrogen Sulfide (H_2S), How to Recognize the Deadly Hazards and Prevent Their Effects*. This book provides details of exposure to hydrogen sulfide and presents multiple case studies of these types of incidents.

When working with or near light naphtha, we should follow precautions or procedures very similar to those for gasoline (petrol in the United Kingdom and elsewhere). Light naphtha is not only highly flammable but can also be explosive under specific air and vapor mixtures.

CSB Investigation Findings

Note that the CSB conducted a very detailed investigation, and many lessons can be learned from this incident. I encourage readers to use the link in the "Additional References" section and review this report thoroughly. I emphasize that what is included here is a summary meant to motivate readers to consult the full CSB final report.

- The CSB determined that the cause of the incident was operators opening valves and removing a flange on the pressurized fuel gas mix drum, which released a flammable liquid, naphtha, directly onto the ground. Once on the ground, the flammable liquid formed a vapor cloud that reached a nearby ignition source, causing a flash fire.
- Contributing to the incident were (from the CSB final report):
  - The refinery's failure to implement effective preventive safeguards for the overflow of towers and vessels in various pieces of equipment, which led to an over-reliance on human intervention to prevent incidents.
  - The refinery's failure to implement a shutdown or hot circulation through the use of stop work authority or otherwise.
  - The refinery's ineffective policies, procedures, and practices to avoid and control abnormal situations.
  - The refinery's alarm system, which flooded operators with alarms throughout the day, resulted in poor decision-making.
  - The refinery's failure to learn from previous incidents.
- The CSB found that in some cases, the BP Toledo Refinery relied on human intervention to respond to process upsets and deviations. Despite the BP Toledo Refinery's reliance on human intervention as an identified safeguard for overflow of the absorber stripper tower to the fuel gas mix drum, the refinery did not adequately consider potential hazards that could exist if the drum contained high levels of flammable liquid (such as naphtha) and needed to be drained.

- The BP Toledo Refinery did not have procedures, written instructions, or documented corrective actions for operators to respond to or troubleshoot a high liquid level in the fuel gas mix drum during either normal operations or process upsets, if liquid entered the drum.
- Furnace safety instrumented systems and emergency pressure-relief valves were also identified as safeguards in the BP Toledo Refinery's PHAs for overflow of the absorber stripper tower to the fuel gas mix drum, but neither was effective in preventing liquid overflow into the fuel gas system.
- Additionally, the industry lacks sufficient guidance on protective systems for a fuel gas mix drum, despite it being an integral part of a refinery's fuel gas system.

Key Lessons Learned (Author's Comments)

The Importance of Implementing an Effective Process Safety and Integrity Management System (PSIMS)

This tragic incident shows a complete failure of the company's and facility's Process Safety Management (PSM) Program. Employees must be properly trained, and company policies should be clear to ensure employees NEVER attempt to drain liquid hydrocarbons directly onto the ground for any reason, especially not to prevent a process unit from shutting down or changing its operation.

Since we were not present during this incident, it is difficult to understand the operators' thoughts and the reasoning behind their decisions. However, I find it hard to comprehend why the operators would deliberately drain light liquid naphtha, an extremely flammable hydrocarbon, directly onto the ground in the presence of ignition sources, such as nearby operating furnaces. I believe they felt immense pressure to keep the units running and avoid an emergency shutdown, thinking they were doing the right thing at the time. This also reflects the organizational culture that existed at the time of the incident. Questions that could be asked include: What is the organization's priority, and how is this communicated to everyone?

It's easy to claim that safety is our top priority, but truly embodying this when immediate decisions are needed – decisions that can mean life or death – is an entirely different matter. Prioritizing avoiding an emergency unit shutdown or even a complete refinery shutdown over a life-and-death situation is unthinkable. Yet, I believe that is exactly what happened here. Draining light naphtha onto the ground, as occurred at this site, endangers everyone present.

Most modern refineries employ a PSIMS to oversee all safety factors in refining and petrochemical operations, and various elements must be controlled. This system includes detailed procedures and practices that ensure safety is incorporated into nearly all decisions and operations at the facility. Process safety fundamentally differs from personnel safety. While personnel safety focuses on preventing slips, trips, falls, and similar incidents or injuries that usually affect only one person or a small group, process safety aims to prevent the release of flammable or toxic materials, which can endanger many people and damage or destroy equipment both within the organization and in surrounding communities and the environment.

This addresses the core of the company's and facility's safety culture. The organization's culture should motivate all personnel to continually watch out for their peers and others, helping to prevent mistakes like the one that occurred here.

Facilities should be designed with detailed step-by-step operational procedures to prevent liquid overfill of process vessels. If there is concern about potential overfill, the appropriate response should be specified in the refinery emergency operations procedures. The correct response should never be to drain light naphtha onto the pavement or ground.

The key elements of the PSIMS primarily involve all factors related to preventing a loss-of-containment incident. Most of these management systems include roughly 11–15 main elements and specify the requirements for documenting each standard and substandard, the responsibilities of each party, and internal and external audits to ensure compliance and ongoing improvement. This includes the "What" (Defines the Scope/Objectives of the system), the "How" (Provides the Procedures/Practices to implement it), the "Who" (Identifies those responsible and accountable for implementation), the "How Do We Know?" (Checks if the expected results are being achieved), and the "How Can We Do Better" (Promotes continuous improvement). Additionally, there are requirements for developing and tracking both leading and lagging indicators to verify performance and ensure that compliance is reflected in the tracking data.

Examples of common PSIMS are listed below (Note that some of these may be combined within certain PSIMS).

The Key Elements of a PSIMS

- Management and leadership commitment and involvement
- Employee participation
- Process safety information
- Required procedures and other documentation
- Training requirements
- Ensuring equipment integrity
- Work permits
- Management of change (MOC)
- Managing contractors
- Prestart-up audits
- Incident and near-miss investigations
- Emergency planning and response

- Internal and external compliance audits
- Trade secrets

The CSB Report did not mention a process integrity management system. They had a stop work authority; however, the CSB stated that this is not a substitute for effective process safety management systems (PSMS).

In my first Wiley textbook, *Process Operations Safety: The What, Why, and How Behind Safe Petrochemical Plant Operations*, I explore the PSIMS, explaining what it is and emphasizing the key elements that should be included. I also detail how implementing such a system can enhance process safety compliance and reduce the history of process safety incidents. I have personally observed how this can happen and the positive effect it has on employee and contractor process safety awareness, which in turn helps prevent loss of containment incidents.

The Importance of an Effective Alarm Management Process

Regarding the alarm overload, the refinery should promptly conduct a comprehensive review of the alarm status and initiate a process to implement an Alarm Management and Rationalization initiative. Rationalization involves removing or relocating low-priority alarms, while other parts of the program focus on fixing alarms that are not working properly or are repeat offenders.

Alarm management includes other improvements, such as ensuring that remaining high-priority alarms have a distinctive audible and visual indication to alert the console operator. This is a major initiative that requires resources from operations and technical teams. However, many consultants and companies can help make the alarms more responsive and efficient.

Draining Hydrocarbons to Grade

Operators should be trained to prevent hydrocarbons from draining to the grade. In this incident, the operators should have recognized that hydrocarbons had accumulated in the fuel mixing drum and that they were highly flammable, with ignition sources nearby. This type of training should be included in operator training and certification.

Process Safety Management System

This incident shows how an effective PSMS backed by strong management support can prevent similar events. This is accomplished through well-documented procedures that are strictly followed by facility management. The PSMS should include both leading and lagging process safety indicators to promote ongoing improvement and should undergo internal and external reviews to track compliance and performance.

Alarm Management

Petroleum refineries and petrochemical plants must ensure their alarm systems work correctly to reduce nuisance alarms, alarm floods, and persistent alarms. Regular reviews of alarm systems should be performed to verify proper alarm priorities. This is usually achieved through a comprehensive alarm management process that includes rationalization, upgrades, and change management for all alarm systems and components.

Operating Procedures

The CSB cited compliance with procedures in their report. Critical procedures, such as operating the fuel gas mixing drum, should be documented in checklists with operator signatures as each step is completed. Procedures should also include Cautions and Warnings to alert operators before performing high-risk steps that could impact safety.

Additional References

M. Darryl Yoes, Hazards of Hydrogen Sulfide: How to Recognize the Deadly Hazards and Prevent Their Effects.

M. Darryl Yoes, Process Operations Safety: The What, Why, and How Behind Safe Petrochemical Plant Operations, Wiley Publishing.

Example Safety Data Sheet for Naphtha, Material Safety Data Sheet: Naphtha, https://www.collectioncare.org/MSDS/naphthamsds.pdf.

Bill Hollifield and Eddie Habibi, The Alarm Management Handbook, Wiley Publishing.

US Chemical Safety and Hazard Investigation Board (CSB), Fatal Naphtha Release and Fire at BP-Husky Toledo Refinery, Oregon, Ohio | Incident Date: September 20, 2022 | No. 2022-01-I-OH, https://www.csb.gov/assets/1/6/final_report_-_20241.pdf.

US Occupational Safety and Health Administration (OSHA), 29 CFR 1910.119, Process Safety Management of Highly Hazardous Chemicals.

Incident 2: Loss of Containment Due to the Failure to Identify Hazards (Hazard Identification)

Imperial Oil Refinery, Sarnia, Canada

Ignition of fractionator internals and collapse of the fractionation column shortly after it was opened for maintenance. The fractionator was destroyed, and no injuries were reported.

April 2, 2019; there were no fatalities or serious injuries

At the Sarnia Refinery, the Mixed Feed Hydrotreater unit was shut down for a planned turnaround on April 2, 2019. The 165-foot-tall fractionation column was prepared by operations for opening using the standard water washing, steaming, and chemical cleaning prior to the installation of isolation blinds and removal of the column access plates. Shortly after opening the column, smoke was observed coming from the column manways just before the column collapsed and fell. The refinery manager was quoted in the local media as saying that previous design changes allowed pyrophorics to heat up and melt portions of the structure. He went on to say that pyrophorics were not adequately considered when the design changes were made.

It is obvious that the vessel was in a sour service, with a product that had a relatively high percentage of hydrogen sulfide (H_2S), which is a precursor to pyrophoric material. The sulfur compounds combine with the iron scale from normal corrosion inside the vessel, creating iron sulfide scale, which is highly pyrophoric. The iron sulfide scale is deposited along the inner vessel walls and on and in the vessel internals, such as the trays, demister screens, internal packing, etc.

The pyrophoric material lies dormant while the vessel is in operation, and one may never even suspect that it is present, that is, until the column is opened to the atmosphere. As soon as the vessel is opened to the atmosphere, air enters the vessel for the first time in several years and combines with the iron sulfide scale. When both top and bottom access plates are removed, the column creates a chimney effect and drafts a high volume of air into the vessel. When the air comes into contact with the iron sulfide scale, the resulting reaction produces extreme heat. This reaction can take place almost immediately, or it may be delayed for hours and sometimes for days. However, when the reaction does occur, the pyrophoric material can ignite, resulting in a smoldering fire inside the process vessel.

While this can be a problem, the real problem is that the smoldering iron sulfide scale can result in the ignition of other combustible or flammable material that may be inside the vessel. This can be residual hydrocarbons, hydrocarbon sludge, rags, scaffold boards, or other similar combustible material. Even worse, the smoldering iron sulfide scale can ignite some of the metal vessel internals, such as demister screens or internal packing. Internal packing, for example, is very dense and may have the iron sulfide scale deeply embedded in the metal components. A metal fire inside these components burns extremely hot and, due to the dense nature of the components, is extremely difficult to extinguish. In these cases, the internal fire dramatically increases the temperature inside the vessel and weakens the tower shell and support structures. This can cause the column to collapse, as happened in Sarnia. See Figure 11.4 for a photograph of the collapsed fractionator at the Imperial Oil Refinery in Sarnia, Canada.

Unfortunately, our industry has a long history of exactly these kinds of events, events where internal pyrophoric fires have resulted in damage to the column shell,

Courtesy of Dave Dentinger, Blackburn Media, Sarnia.

Figure 11.4 Photo of collapsed 150-foot-tall (45.7 m), 30-foot-diameter (9.1 m) fractionation column at the Imperial Oil Refinery, Sarnia (April 2, 2019).

internal damage, or the complete collapse of the fractionation column. Internal fires have also occurred in different kinds of process vessels in addition to fractionators. They have occurred in storage tanks, especially those in our naphtha service, naphtha with high concentrations of hydrogen sulfide, on various kinds of fractionation equipment, on sour gas compressors, and other equipment in sour service. It is noteworthy that, in many of these instances, there was no history of previous incidents with pyrophoric materials, or at least no one at the site knew of the prior incidents.

For example, this same site reportedly experienced a similar incident in 1988 when an MEA Column was opened. In that case, after tower steaming and blinding, manholes were opened on an MEA Absorber. Several hours later, iron sulfide on the internal random packing materials ignited, causing internal tower damage.

In 2003, another pyrophoric fire occurred at an Asian Refinery when a large floating roof tank in sour naphtha service experienced an internal fire. The tank was opened for cleaning, inspection, and maintenance, and air eductors were placed on the manways to assist in gas-freeing the vessel. Shortly after the manways were removed, a smoldering fire ignited on the interior walls of the tank shell and the underside of the floating roof. This resulted in several explosions when the smoldering pyrophoric embers fell into the residual light naphtha. The internal smoldering and subsequent explosions happened shortly after the manways were taken out, and work had begun earlier that day to vacuum the remaining residual naphtha from the tank. Fortunately, no one was injured; however, the paint on the tank shell was damaged, and the gauging pole needed replacement.

Several good lessons were learned here, and as a result of this incident, the facility has developed enhanced plans to prevent ignition by pyrophoric materials. These plans include continuous wetting and steaming of tank internals during cleaning and potential neutralization of pyrophoric iron sulfide scales through chemical cleaning before the vessel is opened.

There is also a plan to identify other equipment that may be exposed to pyrophoric materials and to develop more specific plans for addressing pyrophoric materials during equipment opening. Additionally, there are plans to enhance and formalize firefighting techniques for preventing and managing internal pyrophoric fires.

Another incident similar to the Sarna tower collapse occurred at the Chevron Phillips Chemical Plant in 2001. In this case, the unit was undergoing maintenance, and welding on the internals was in progress in the styrene monomer unit fractionator, a 250-foot (76.2 m) tall and 30-foot (9.1 m) diameter process vessel. It is believed that the welding inside the column accidentally ignited the metal packing within the fractionator, although ignition by pyrophoric materials can't be ruled out.

Once ignited, these fires are extremely difficult to extinguish. Water is not particularly effective due to the extremely high temperature of a metal fire. Metal fires have been known to burn so intensely that water is converted to its elements (hydrogen and oxygen), resulting in an even larger fire and potential explosions due to the presence of hydrogen. For these reasons, we recommend that internals, especially demister screens and tower packing, be removed from the section of the column where hot work is planned. Of course, this same incident can occur because of pyrophoric materials that ignite the metal internals (demister screens and internal tower packing).

Another example of a pyrophoric ignition happened at the 160,000 bbl/day Citgo Refinery in Lemont, Illinois, on August 14, 2001. This refinery had a major fire in the atmospheric crude unit because a piping elbow made from a different material than specified failed (a positive materials identification failure [PMI]).

The fire damaged the piping systems around the crude unit, affecting several other pipelines. Several days after the fire was extinguished, air was drawn into the crude atmospheric column through open pipes that ruptured during the incident. This led to a pyrophoric ignition inside the column, causing the crude tower to collapse. Due to the extensive damage from the fire and collapse, the crude distillation unit remained out of operation for about a year.

In addition to the above pyrophoric incidents and those following, I am aware of at least eight other similar incidents, most involving ignition of tower metal internals. Several of these are briefly discussed below:

- Valero Refinery, Wilmington, California (January 2004):
  At the start of the FCCU turnaround, the FCCU fractionator was steamed for 28 hours, chemically cleaned for more than 36 hours, and rinsed for more than 30 hours. Fourteen hours after the manways were opened and air movers installed, a fire was discovered in the structured packing. The fire was

eventually extinguished with water; however, the column shell was distorted, and 2 packing beds were damaged. Iron sulfide deposits were embedded in the metal packing and ignited when contacted with air.

- ExxonMobil Chemical Plant, Baton Rouge, Louisiana (January 2001):
  During a plant turnaround, a high-temperature metal fire occurred in the titanium packing beds in a chemical fractionator after the access plates were removed. It is believed that the titanium packing was ignited by the pyrophoric catalyst, which had carried over from the reactor. The fire very badly damaged the column shell and internals.
- BASF Chemical Plant, Port Arthur, Texas (September 5, 2023):
  In preparation for a maintenance turnaround, a few weeks before the incident, the refinery took the Pyrolysis Gasoline unit offline, including the Extractive Distillation Tower. The plant followed its normal procedure to remove solvent and hydrocarbons from the column and prepared to open it for entry and maintenance. On the morning of the incident, a maintenance crew began removing the manway plates to prepare the vessel for entry. Around 9:30 a.m., internal temperatures within the column's stainless steel structured packing beds started to rise. By 10:44 a.m., visible smoke and flames erupted from the vessel, clearly indicating an internal fire. The emergency alarm was activated, and all personnel evacuated the area. Nitrogen was injected into the column in an attempt to extinguish the internal fires. Around 11:53 a.m., the tower collapsed. See Figure 11.5 for a photograph of the collapsed Port Arthur, Texas, BASF Total Pyrolysis Gasoline Extractive Tower.

In my first Wiley textbook, *Process Operations Safety: The What, Why, and How Behind Safe Petroleum Plant Operations*, I covered the hazards of pyrophorics in detail. Additional information and incident reviews are available there.

Key Lessons Learned

The Basics

- Internal demister screen and packing material are effective at fractionation due to the extremely high surface area relative to the volume ratio. However, this means that there are many small openings and crevices that can accumulate iron sulfide scale. This is not a problem when the column is in service, but due to the pyrophoric nature of the material, it is a serious fire hazard when the vessels are opened to the atmosphere, and air is allowed into the vessel.
- When an internal fire does occur, it is extremely difficult to extinguish. Metals burn at very high temperatures. If water is used as the firefighting agent, the high-temperature thermite reaction can separate the hydrogen from the oxygen, essentially adding fuel to the fire.

US Chemical Safety and Hazard Investigation Board (CSB). Public Domain.

Figure 11.5 Photograph of the damaged BASF TOTAL Energies Petrochemical, LLC ("BTP") facility in Port Arthur, Texas. The column collapsed due to a pyrophoric fire inside the distillation tower, which ignited internal metal components, weakening the column shell.

- Due to its compactness, internal packing is extremely difficult to clean and even more difficult to inspect for cleanliness.
- When a column is open to the atmosphere, especially if the lower and upper manways are opened, there is a chimney effect that draws fresh air into the column, setting the stage for an internal fire.
- Some precautions can and should be taken; however, we should recognize that they may not always be effective. This is why we will talk about fire prevention and protection in more detail below.

Equipment Preparation for Opening

Let's begin by preparing the column for opening, entry, and maintenance. Most vessels that have experienced internal fires after being opened have been cleaned using steaming, water washing, and chemical cleaning. This is an effective method for preparing a column, unless there is a risk of pyrophoric iron sulfide scale or other pyrophoric deposits. Chemical cleaning is an additional step typically performed with a chemical de-oiler or a similar product. This procedure cleans the interior of oil or hydrocarbon residue, but unfortunately, it is not very effective at neutralizing pyrophoric material.

I am familiar with sites that have successfully neutralized pyrophoric material by circulating a mild solution of potassium permanganate, typically around 1% in water. This has been done safely many times; however, a word of caution when using potassium permanganate: it should never be used in a low pH environment, as this may result in an undesirable reaction of the permanganate. I advise always consulting with a reputable chemical cleaning vendor and a chemical engineer before using potassium permanganate. What I have seen is that the process begins with circulating the 1% solution, which is purple. The circulated material will quickly change to a brown color, and at this point, the chemical vendor will slowly add the chemical until the color returns to purple. When the circulated chemical changes back to purple, this indicates that the neutralization is complete, and the circulation is discontinued. There are commercially available chemicals that advertise the neutralization of pyrophorics; however, the active ingredient is proprietary. I suspect that, at least for some, the active ingredient may be permanganate.

Fire Prevention and Protection Precautions

Fire protection precautions must be established for columns, tanks, or other process equipment in sour service, which has a relatively high concentration of hydrogen sulfide or other sulfur compounds, before any access plates or openings are opened. I always recommend positioning a charged fire hose at each manway or access plate before opening. Once the access plate is opened, a brief burst of fire water should be released to moisten the vessel's interior. Flooding the vessel with water or maintaining a continuous stream is typically unnecessary. Our goal is to prevent the drying of the vessel internals, which can lead to the ignition of pyrophoric materials. This wetting action should occur periodically until the vessel is entered and the residual sludge or scale is removed. Due to the higher-pressure requirements, a fire water booster pump may be necessary to access the upper entry points for taller process vessels.

Fire protection should involve practices that prevent hot work in columns and other vessels equipped with demister screens or internal structured or random packing. As demonstrated in the incidents above, this material ignites easily and is difficult to extinguish. Maintenance practices should include removing the demister screen or packing materials in areas of the column where hot work is planned. The remaining sections containing demisters or packing should also be sealed with insulation or fire blankets to prevent sparks from entering these areas. An even better alternative might be to periodically wet the adjacent areas to avoid drying and ignition of the metal components.

It may also be a good idea to use a carbon monoxide (CO) detector inside the vessel when it is open to the atmosphere. This can provide an early warning of a potential fire or smoldering in the column that may be otherwise undetectable.

Additional References

Asphalt Institute, Pyrophoric Material Formation in Heated Asphalt Storage Tanks, https://www.asphaltinstitute.org/wp-content/uploads/AI-Fire-Safety_rev2.pdf.

The Associated Press – SouthCoastToday, Jury Awards $387.4 Million to Citgo to Cover Costs in 2001 Fire at Illinois Oil Refinery, https://www.southcoasttoday.com/story/business/2006/11/08/jury-awards-387-4-million/50375317007/.

Carus (Carus' CAIROX® Potassium Permanganate and CARUSOL® Sodium Permanganate), Pyrophoric Iron Sulfide Removal, https://www.carusllc.com/industries/oil-and-gas/.

CBC, Pyrophoric Materials Caused Imperial Sarnia Tower to Fall, https://www.cbc.ca/news/canada/windsor/imperial-sarnia-tower-pyrophoric-1.5164973.

Chron, Citgo Wins Millions in Refinery Fire Lawsuit, https://www.chron.com/business/energy/article/citgo-wins-millions-in-refinery-fire-lawsuit-1895310.php.

Global News, Tower Falls at Imperial Oil Site in Sarnia, 'High Flaring' Seen from the Site: Company Statement, https://globalnews.ca/news/5124964/tower-fall-imperial-oil-sarnia/.

The Journal, Design Flaw, Byproduct Caused Tower Collapse: Imperial, https://www.thesarniajournal.ca/top-story/design-flaw-byproduct-caused-tower-collapse-imperial-7969470.

Oil and Gas Journal, Improved Cleaning Method Safely Removes Pyrophoric Iron Sulfide, https://www.ogj.com/home/article/17242644/improved-cleaning-method-safely-removes-pyrophoric-iron-sulfide.

Sarnia News Today, 'Design Change' Contributing Factor in Imperial Tower Collapse, https://sarnianewstoday.ca/sarnia/news/2019/6/6/design-change-contributing-factor-imperial-tower-collapse.

West Penetone (https://www.westpenetone.com), Potassium Permanganate Solution - Iron Sulfide Particulate Remediation, https://www.penetone.com/wp-content/uploads/2017/08/Permanganate_Solution_EN-1.pdf.

Work Safe BC, Pyrophoric Materials Create Fire and Explosion Hazard, https://www.worksafebc.com/en/resources/health-safety/hazard-alerts/pyrophoric-materials-fire-explosion-hazard?lang=en.

Incident 3: Loss of Containment Due to the Failure to Identify Hazards (Hazard Identification)

ECO Special Waste Management Pte Ltd ("ECO SEM"), Singapore

A fire occurred while transferring a nonconductive product to a plastic container and quickly spread site-wide due to the presence of flammable liquids. No injuries were reported, although the facility was "badly damaged."

February 23, 2017; no injuries were reported

A flash fire occurred at this waste management facility when hexane was being drained from a product tank through a metal hose into a plastic, nonconductive intermediate bulk container. When the flash fire happened, it quickly spread due to the presence of other flammable liquids and vapors. The local fire rapidly spread to other parts of the facility and across the site. Although the workers tried to extinguish and control the initial fire, it quickly expanded.

The good news is that all workers safely evacuated the facility, and the Singapore Civil Defence (Defense) Force controlled and extinguished the fire. No workers were hurt, but the Ministry of Manpower (MOM) reported that the facility was extensively damaged by the fire.

The following quotes are from the MOM regarding this incident:

> "Investigations by the Ministry of Manpower (MOM) revealed that ECO SWM failed to conduct a risk assessment to ascertain the suitability of storing highly flammable liquid Hexane in plastic IBCs. The IBC used by the ECO SWM was not suitable as it could not dissipate the electrostatic charges generated during the discharging of Hexane, which ignited the flammable Hexane vapour (vapor) – air mixture that resulted in the fire."
>
> "The company also failed to ensure that its workers adhered to safe operating procedures for dealing with hazardous products. Hexane and another flammable liquid, Xylene, were spilled onto the floor during previous work processes, and were not cleaned up immediately by the workers. The spilled liquids formed flammable vapours (vapors) in the area, which fueled the spread of the fire."

Er. Go Heng Huat, The MOM's Director of the Major Hazards Department, said:

> "Companies in hazardous waste treatment business must ensure that a comprehensive risk assessment of their work processes is conducted and all necessary risk control measures are implemented. They also have the duty to ensure that workers understand and adhere to safe operating procedures at all times. MOM will not hesitate to prosecute companies that disregard hazards at the workplaces and contravene safety regulations."

Key Lessons Learned (Author's Comments)

- This company chose to transfer a nonconductive liquid into a plastic storage container not designed for the storage of this product. The key learning is that nonconductive liquids are static accumulators, and nonconductive plastic containers cannot be grounded to discharge the accumulated static charge.

- The other key learning is that naphthas were allowed to accumulate throughout the facility and quickly ignited due to the relatively minor fire that occurred during the transfer of product. The initial fire quickly ignited the accumulated hydrocarbons, making this a site-wide emergency and causing significant damage to the facility.

Static Electricity References

Many petroleum products are nonconductive and can therefore generate significant static electricity. Hexane, although not technically classified as naphtha, is derived from naphtha and falls into this category in terms of electrical conductivity. Many other petroleum products also belong to this category. Examples include the following (not intended to be an exhaustive list):

- Jet fuel
- Solvents
- Diesel (especially ULSD)
- Naphtha
- Gasoline
- Hexane
- Benzene
- Toluene
- Xylenes
- Some lube base stocks

The amount of static buildup in liquids is directly related to their electrical conductivity. Conductive liquids usually do not pose significant static hazards because they can quickly dissipate static charges through the liquid to the tank's steel or conveyance, causing the static charge to be rapidly released. Although the following liquids are generally considered conductive, I recommend testing the liquid for conductivity before assuming it is conductive and not a static hazard. Examples of liquids that are usually conductive include the following (not an exhaustive list):

- Black oils
- Crude oils
- Residual fuel oil
- Bitumen/asphalt
- Alcohols, ketones
- Water

Two key references that offer detailed information on handling these products and preventing fires caused by static discharge are the National Fire Protection Association (NFPA) Code 77 and the American Petroleum Institute (API) Recommended Practice RP 2003.

The NFPA code 77 states that "under certain conditions, particularly with liquid hydrocarbons, static may accumulate in the liquid, with the danger of subsequent sparking in a flammable vapor-air mixture."

The API Recommended Practice 2003 "Protection Against Ignitions Arising out of Static, Lightning, and Stray Currents" details methods to prevent hydrocarbon ignition caused by static electricity, lightning, and stray currents. The standards outlined in the API RP are incorporated by reference into the US Code of Federal Regulations (CFR) and are regarded as federal requirements in the United States.

Although not involved in this incident, it is important to note that hydrocarbon mists, aerosols, and dust clouds can also collect static electricity. Catastrophic explosions have happened in the United States and worldwide because of the buildup of combustible dust in the air. Several of these major accidents were triggered by static discharge within the dust cloud.

Examples of combustible dust involved in dust cloud explosions include the following (not meant to be an exhaustive list):

- Coke dust or particulates
- Plastic dust or particulates
- Coal dust or particulates
- Metal dust
- Grain dust (sugar, corn, etc.)
- Wood dust

Key Lessons Learned (Author's Comments)

The incident report available on the MOM's website did not mention two other important elements involved in this investigation. They did not include information about the employees' training related to handling and processing flammable materials, especially the hazards of nonconductive flammable liquids, the types of containers to use, and the requirements for electrical bonding and grounding of these conveyances and storage containers.

Likewise, employees should be trained to stay alert for the buildup of flammable or combustible hydrocarbons on the ground and in open-top storage containers, where they can be easily ignited by minor arcing or even an open fire. This helps prevent the rapid spread of a large fire that could engulf the entire area facility.

The incident report also did not mention whether detailed procedures are available for employees to follow when handling these types of flammable liquids and during product transfer tasks. The facility should have these procedures documented as a step-by-step process, including cautions and warnings to alert employees about critical steps, such as choosing proper storage containers and ensuring electrical bonding and grounding to prevent static discharge.

Additional References

Singapore Ministry of Manpower, Company Fined $230,000 for Unsafe Work Practices that Led to Fire at Hazardous Waste Treatment Facility in Tuas, https://www.mom.gov.sg/newsroom/press-releases/2019/0828-company-fined-for-unsafe-work-practices-that-led-to-fire#:~:text=The%20Ministry%20of%20Manpower%20(MOM)%20investigated%20the,brought%20the%20fire%20under%20control%20at%2011.00am.

The Straits Times, Tuas Fire Put Out After 4 Hours, No Casualties Reported, Cause of Fire Under Investigation: SCDF, https://www.straitstimes.com/singapore/fire-breaks-out-at-tuas-view-circuit-public-advised-to-avoid-the-area.

Today Online Singapore, Huge Fire at Tuas Waste Management Plant, Nearby Roads Closed, https://www.todayonline.com/singapore/fire-tuas-roads-closed-facilitate-operations.

WMAM – Waste Management Association of Malaysia, Huge Fire Breaks Out at Waste Management Plant in Tuas, https://wmam.org/huge-fire-breaks-out-at-waste-management-plant-in-tuas/.

Incident 4: Loss of Containment Due to the Failure to Identify Hazards (Hazard Identification)

ExxonMobil Refinery, Baton Rouge, Louisiana

November 22, 2016; zero fatalities, six injuries, including four seriously burned

On November 22, 2016, an operator was isolating an isobutane line for maintenance by removing the motor operator from an isolation valve to close the valve manually. This resulted in a large isobutane release and fire in the sulfuric acid alkylation unit at the ExxonMobil Refinery, which resulted in four serious injuries to workers and injuries to two others.

The CSB investigated this incident. Below is a summary of the final report. The CSB also released a safety video and a Safety Bulletin regarding this incident. Excerpts from the final report are included below.

On November 22, 2016, an isobutane release and fire seriously injured four workers in the sulfuric acid alkylation unit at the ExxonMobil Refinery in Baton Rouge, Louisiana ("Baton Rouge Refinery"). The alkylation unit had motor-operated isolation valves on most liquified petroleum gas (LPG) pipelines. These were quarter-turn plug-type valves that employed two designs for mounting the motor operators: an older design and a more recent model. On that date, the operator attempted to isolate an isobutane piping system, but the motor operator failed to function. As before, the operator tried to remove the motor operator from the inoperable motor-operated valve, which used the older design, and planned to operate the valve with a pipe wrench by manually rotating the valve plug.

Unfortunately, the older design was mounted so that when the top cap mounting bolts were removed, the valve released directly into the atmosphere, releasing a significant amount of LPG. Refer to Figure 11.6 for a sketch illustrating this design and how the valve was opened to the atmosphere when the motor support bolts were removed. Figure 11.7 shows the location of the bolts that could have been removed to remove the valve motor actuator.

The operator removed the four bolts that secured the motor operator mounting adapter plate and attempted to rotate the valve stem using a pipe wrench. However, as the valve turned, the internal components blew out, resulting in a significant loss of containment of isobutane, which rapidly formed a flammable vapor cloud. The flammable cloud ignited in about 30 seconds, causing severe burns to four workers. Unfortunately, the two remaining bolts that could have been used to remove the motor operator remained in place.

Other personnel working on the unit were evacuated quickly, and the console operators were directed to initiate an emergency shutdown of the unit. The CSB believed that the most likely ignition source was a welding machine operating about 70 feet away from the release.

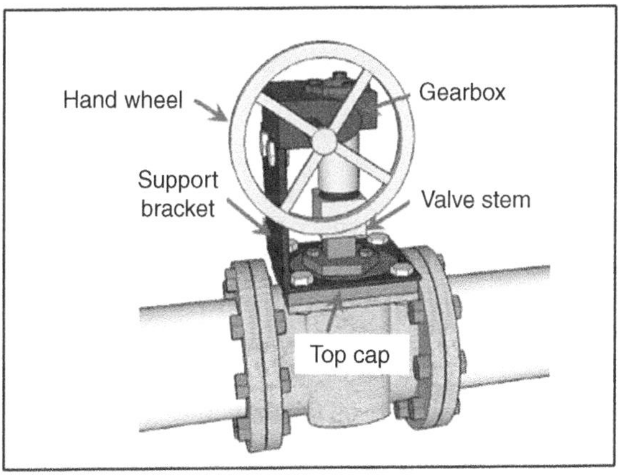

Figure 11.6 Depiction of the ExxonMobil Sulfuric Acid Alkylation Unit isolation valve with older design. Removing the four bolts that secure the valve top cap opens the valve body to the atmosphere. This illustration is taken from the Safety Bulletin on this incident, which was prepared by the CSB. US Chemical Safety and Hazard Investigation Board (CSB). Public Domain. Note that in this incident, the valve was equipped with a motor operator instead of a hand wheel; otherwise, the components of the valve were identical.

US Chemical Safety and Hazard Investigation Board (CSB). Public Domain.

Figure 11.7 Illustration of a similar plug valve highlighting the bolts that could have been removed to safely separate the motor operator from the valve body. Removal of these bolts does not affect the security of the top plate and does not result in the product being released. This image comes from the Safety Bulletin published by the US Chemical Safety and Hazard Investigation Board (CSB), which discusses the lessons learned from this incident.

The CSB Incident Analysis

The CSB found several Baton Rouge Refinery safety management system deficiencies that led to the removal of the plug valve gearbox and the inadvertent disassembly of its pressure-retaining top cap, resulting in an isobutane release and fire.

These deficiencies included:

- Failure to identify and address the older model plug valve design and gearbox reliability issues.
- Lack of a human factors evaluation to identify the older model plug valves' design and reliability issues, as well as the potential hazards associated with operating and maintaining these valves.
- No written procedures detailing the steps needed to remove different models of gearboxes from plug valves to manually open or close the valve safely.

- There was no training for workers to safely remove the various plug valve gearbox models in the alkylation unit and on the hazards associated with this type of work.
- An organizational culture that accepted operators removing malfunctioning plug valve gearboxes despite the lack of detailed procedures and training for safe removal.

Author's Comments: The Value of Conducting a Last-minute Risk Assessment (LMRA)

This incident occurred when the operator attempted to close an isolation valve after the motorized actuator failed to operate. The operator mistakenly removed the wrong bolts, which caused the unintended removal of the valve's top plate and resulted in the loss of containment of LPG, isobutane – a highly volatile, pressurized liquid hydrocarbon. LPG quickly transforms from a liquid into a gas when the pressure drops. For example, propane expands 270 times when it vaporizes from liquid to gas.

I have no issue with the CSB's findings; however, this seems to be a case where the incident might not have happened if the worker had paused to perform a LMRA and asked himself, "What can go wrong? or What can happen if I do this task?"

An LMRA is a quick, personal final check meant to identify potential safety, health, and environmental risks in the workplace before starting a job or task. Its purpose is to find any hazards or safety, health, environmental (SHE) issues that might be present before work begins. This incident caused four employees to suffer serious burns, including the operator who performed this task.

I believe that if the LMRA had been performed, the employee might have realized that these bolts were securing the valve top cap and that removing them could potentially release LPG.

Additional References

British Health and Safety Executive (HSE), Last Minute Risk Analysis (LMRA), https://hselifenl.com/wp-content/uploads/2017/07/HSElife-NL-LMRA-A6-UK-2017.pdf.

Product Safety Data Sheet, Isobutane, https://apps.spheracloud.net/Pdf/MsdsIAImages/ekifYOipskRW8JfqaREQ.PDF.

US Chemical Safety and Hazard Investigation Board, ExxonMobil Baton Rouge Refinery Chemical Release and Fire, https://www.csb.gov/investigations/completed-investigations/?F_InvestigationId=3590.

Incident 5: Loss of Containment Due to Failure to Identify Hazards (Hazard Identification)

Delaware City Refinery, Delaware City, Delaware

A loss of containment and flash fire occurred while draining a piping system of what was believed to be steam condensate.

November 29, 2015; one person was seriously injured, requiring hospitalization

The CSB investigated this incident and provided a detailed report. This report includes a summary and excerpts from the CSB report. You are encouraged to review the CSB report for more details.

Overview

An operator at the Delaware City Refining Company (DCRC) suffered second-degree burns to the face and neck, and third-degree burns to the wrist, on November 29, 2015, when he was performing a nonroutine process task, the draining of steam condensate from a process vessel in preparation for the removal of a piping spool that led to a connected process. Unfortunately, due to failure to fully isolate the system, hydrocarbons had entered the system being drained, and when it was drained to the refinery sewer system, a flash fire occurred resulting in the operator's injuries.

The operator was transported by ambulance to the hospital, where he was treated for burn injuries and was released three days after the incident. This incident was investigated by the CSB, and a detailed investigation report was issued. What is provided here is a summary, and I have condensed the CSB report with the intent to make it easier to quickly interpret and understand. Readers who are interested in understanding the details of this incident are encouraged to access the full CSB report using the link provided in the "Additional References" section of this chapter.

The CSB reported that this incident follows two other incidents at the same facility, which occurred on August 21 and August 28, 2015, and of even more significant concern is the number of incidents like this that the CSB has investigated. The CSB reported that of all the incidents the CSB has investigated, 37% occurred in chemical and manufacturing facilities either prior to, during, or immediately following maintenance activities. These are primarily OPE types of incidents! Also of major significance, these incidents resulted in 86 fatalities and 410 injuries.

This supports exactly why this book is being written and distributed. Unfortunately, OPE incidents are almost commonplace in the petroleum refining and petrochemical industries, and many have had tragic outcomes. These are the kinds of events that change lives forever, and they are preventable.

The Basics of Alkylation

In alkylation, olefin hydrocarbons, hydrocarbons with double bonds, are combined with an isoparaffinic, typically isobutane, to create a longer chain molecule that has a much higher octane. Alkylation became a very important technology during World War II due to the requirement for higher octane gasoline to support the war effort with high-octane aviation fuels.

A refinery alkylation unit produces the high-octane gasoline components known as alkylates, which are blended into motor and aviation gasolines. The feed, which is primarily olefins, is produced from the FCC unit and primarily butylene, but also propylene and amylene, along with isobutane to help with the reaction. Alkylation feed maintains a high isobutane to olefin ratio to aid the reaction producing gasoline range material (alkylates) from olefins such as propylene, butylene, and amylene. Butylene is the most sought-after feedstock due to the high octane alkylate produced.

The petroleum refining alkylation reaction occurs in a stirred bed reactor with an acid, either sulfuric acid or hydrofluoric acid, as the catalyst. The alkylation systems operate under pressure to ensure that the light hydrocarbons remain in a liquid state. Also, due to the highly exothermic reaction that occurs in the reactor, and to ensure the selectivity of the reaction, the process also uses an auto-refrigeration process to ensure that a low temperature is maintained in the reaction bed. The products from the reactor are then routed to an acid settler vessel to separate the acid from the hydrocarbon products. The reactor products are also treated in the neutralization section to remove or neutralize any trace amounts of acid before being fractionated to separate the alkylate from the propane and butane streams, and all products are sent to refinery tankage as gasoline components, and as LPG products.

HF is a highly toxic material with a very low boiling point and has been the subject of serious incidents involving alkylation at other sites in the United States. Due to its toxicity and low boiling point, HF can become a highly toxic fuming vapor when released to the atmosphere. HF is not used at the Delaware Refinery and did not play any role in this incident.

The alkylation unit at the Delaware Refinery has two different alkylation technologies that make up the alkylation unit, with both technologies using sulfuric acid (H_2SO_4) as the catalyst. One unit uses a stirred bed reactor, and the other is the STRATCO process. The primary difference in the two technologies is the design of the reactor and how the olefin feeds are contacted with the catalyst.

This incident occurred in the neutralization section, where trace amounts of acid are neutralized by "scrubbing" with caustic (sodium hydroxide – NaOH) before being sent to the fractionation columns. The neutralization section features a mixer, a caustic settler, and a coalescer vessel. The neutralization section removes the remaining acid or acid products by contacting the incoming reactor effluent stream with fresh caustic in an on-stream mixer. The fully mixed stream flows into the caustic settler, where the caustic and any remaining contaminants are settled to the bottom of the vessel and drawn off, allowing the "purified" product stream to flow to the coalescer. The coalescer removes any remaining trace amounts of caustics or other contaminants allowing the cleaned stream to flow to the downstream fractionation columns, in this case, a depropanizer column. Any trace amounts of acid carried over from the reactor are neutralized in this process.

See Figure 11.8 for a sketch, courtesy of the CSB, that illustrates a simplified flow for the caustic neutralization section and illustrates the section of piping that was involved in this incident.

Equipment Isolation

A 20-foot (6.1 m) section of light hydrocarbon pipe in the neutralization area was previously found to be internally corroded and was scheduled for replacement. A plan was developed to isolate the pipe segment, fill it with condensate (water), and drain it into the oily water sewer to facilitate the replacement.

However, during the isolation process, the crew discovered that one of the isolation valves was leaking internally. As a result, the isolation plan had to be expanded. This expanded plan included additional process equipment and extended the isolation to a larger network of piping.

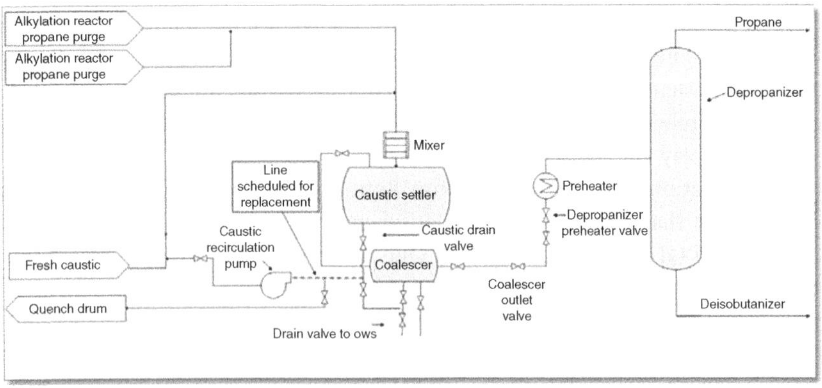

Figure 11.8 A simplified flow schematic of the Alkylation caustic neutralization section. US Chemical Safety and Hazard Investigation Board (CSB). Public Domain.
Note: Annotations are by the US CSB.

The day shift finished the isolation, and the equipment was handed over to the night operator to complete the process by draining the isolated piping circuit in preparation for piping removal by the maintenance team.

However, the expanded isolation plan still included a single isolation block valve, different from the one identified earlier. This valve isolated the line being drained from other process equipment that remained filled with highly volatile and flammable light hydrocarbons, specifically isobutane and propane. The interconnected systems were filled with light hydrocarbons and maintained under process pressure.

The CSB found the expanded isolation system to be more complex and nonroutine, which the operators confirmed. Operators told the CSB that they had never seen isolation performed in this manner.

Unnoticed by the night operator, the single isolation valve in the expanded isolation plan was also leaking internally, which allowed light hydrocarbons to escape into the circuit that had just been isolated for draining. This displaced some of the water used to clear the system of hydrocarbons and partially filled the system with the light hydrocarbon product.

See Figure 11.9, which illustrates the original isolation strategy; however, due to one of the valves leaking through, the isolation was expanded to include additional equipment.

Figure 11.10 shows the expanded isolation plan and the additional equipment that was isolated. Unknown to the operators, a second valve also leaked internally, leading to this incident.

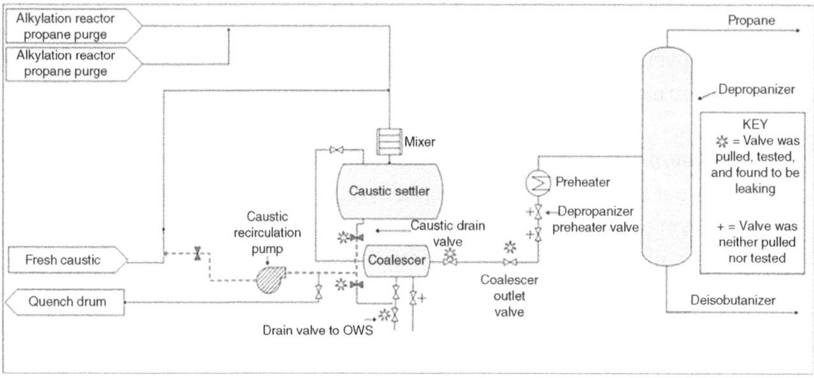

Figure 11.9 The neutralization section illustrates the initial isolation strategy and is from the CSB investigation report. The leaking valve necessitated expanding the isolation plan, which added more equipment to separate this part of the unit from the operational equipment. The isolated equipment is shown with dotted lines. US Chemical Safety and Hazard Investigation Board (CSB). Public Domain.

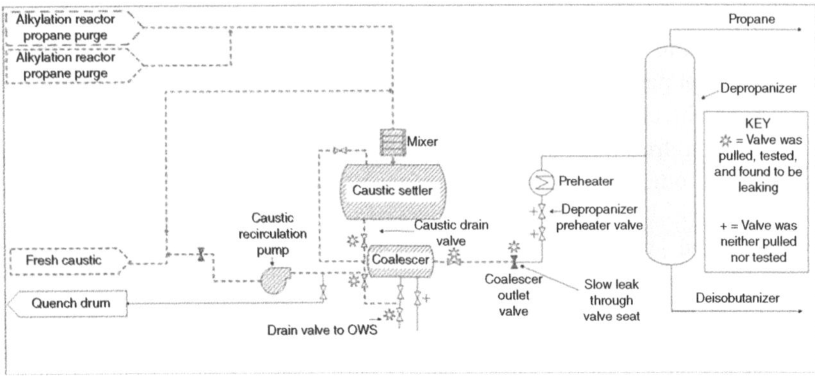

Figure 11.10 This drawing shows the expanded isolation strategy, the amount of additional equipment involved, and the second valve that was determined to leak internally, leading to this incident. Equipment isolation is shown with the dotted lines. US Chemical Safety and Hazard Investigation Board (CSB). Public Domain.

The Incident

When the night operator opened the vessel drain valve to drain what he thought was steam condensate, he inadvertently released the light hydrocarbons that had backed into the isolated system through the leaking single isolation valve. The light hydrocarbons were released into the oily water sewer system, where they quickly vaporized, allowing the volatile vapor to flow upward to the process areas through the pump pad drains.

The purpose of the pump pad drains is to drain any rainwater or other liquids from the pump pads into the oily water sewer. However, in this case, it created a direct path from the oily water sewer to the atmosphere. Unfortunately, a furnace was operating very near one of the pump pads, and the CSB believes the vapors reached this furnace and were quickly ignited, resulting in the flash fire.

Key Lessons Learned (Extracted from the CSB Report)

- Operational tasks that involve preparing equipment for maintenance can be uncommon and nonroutine; therefore, ensure procedures include steps for preparing all process equipment for maintenance.
- For all equipment preparation activities, develop a process that requires preplanning and hazard identification prior to initiating the work, which includes assessing hazards that may be introduced and steps for mitigating those hazards.
- When isolating equipment for de-inventorying or decontaminating activities prior to maintenance work, avoid relying on single block valves, even when the valves are newly installed, to control the release of hazardous energy.

- When an equipment preparation task or isolation plan needs to be modified or expanded due to leaking valves or changing conditions, initiate a process to evaluate hazards that may be introduced by the change.
- Avoid draining materials containing, or potentially containing, hydrocarbons or flammables into the sewer or other systems not specifically designed for such services, especially when ignition sources are present.

Author's Additional Thoughts for System Isolation

When isolating equipment, including piping systems, for opening, always consider using a double-block and bleed isolation process when available. To achieve zero energy, the bleeder valve between the double-block and bleed valves should be fully open and maintained in a car-sealed open position before issuing the work permit. If double block and bleed valves are not available, a risk assessment with an elevated level of review should be conducted to better understand the hazards involved and how to mitigate them.

When preparing equipment for opening, such as when steaming or draining a pipeline, a pump, or other process equipment, it is best to use temporary hoses and vent the equipment to a closed system like a flare or a slop system. This prevents the release of hydrocarbon liquids or vapors into the atmosphere where ignition sources may be present.

This incident caused the operator to suffer severe burns and require several days of hospital treatment. The operating crew was fortunate that this incident did not result in more serious consequences. A loss of containment of LPG or light naphtha in the presence of multiple ignition sources can, and has, led to explosions with catastrophic effects on personnel and significant damage to equipment.

These kinds of incidents also often lead to negative impacts on the company's reputation and significant legal consequences as well. This emphasizes the main points made by the CSB in their report, summarized above in the "Key Lessons Learned."

This primarily involves ensuring equipment is isolated and verified to a zero-energy state before opening drains or handing over equipment to craftsmen for maintenance. Operators must be trained to check vents and drains and to confirm that equipment is free of hydrocarbons and toxics before opening drains, flanges, or access plates.

I have found that the best approach is to use checklists that break tasks down into individual steps, listed in the order in which each step should be completed. Operators should be expected to sign off on each step as it is completed. This helps prevent assumptions by the operator and ensures that steps are finished, not skipped or missed, and are completed in the right sequence. It also helps ensure accountability for the tasks being performed.

Operators may oppose this, and most believe it's unnecessary. They think the procedures are simple and straightforward, and to some extent, this is true. However, many of these individual steps are critical, and missing one step, making an omission, or having an incorrect assumption can lead to serious issues. After all, saving one life or preventing a serious incident makes the extra effort worthwhile.

Additional References

Cameo Chemical Database, Caustic Soda Solution, https://cameochemicals.noaa.gov/chemical/1499.

Cameo Chemical Database, Sulfuric Acid, https://cameochemicals.noaa.gov/chemical/5193.

Center for Chemical Process Safety (CCPS), Line Opening – Strategies & Effective Practices to Manage and Mitigate Hazards, https://www.aiche.org/ccps/resources/tools/safe-work-practices/line-opening/strategies-effective-practices-manage-and-mitigate-hazards.

ExxonMobil, Opening Process Equipment, Important Considerations, https://refiningcommunity.com/wp-content/uploads/2017/07/Opening-Process-Equipment-OPE-Important-Considerations-Moloney-ExxonMobil-DCU-FCCU-SRU-Budapest-2017.pdf.

Inspectioneering, CSB Releases Safety Bulletin into 2015 Chemical Release and Flash Fire at the Delaware City Refining Company, https://inspectioneering.com/news/2017-05-18/6497/csb-releases-safety-bulletin-into-2015-chemical-release-and-flas.

US Chemical Safety and Hazard Investigation Board (CSB), CSB Issues Safety Bulletin – 2015 Chemical Release and Flash Fire, Delaware City Refining Company, https://www.csb.gov/csb-releases-safety-bulletin-into-2015-chemical-release-and-flash-fire-at-the-delaware-city-refining-company/.

US Chemical Safety and Hazard Investigation Board (CSB), Delaware City Refining Company, https://www.csb.gov/investigations/completed-investigations/?F_InvestigationId=3584.

US Chemical Safety and Hazard Investigation Board (CSB), Delaware City Refining Company Investigation Details, https://www.csb.gov/delaware-city-refining-company/.

US Chemical Safety and Hazard Investigation Board (CSB), Removal of Hazardous Material from Piping Systems, https://www.csb.gov/removal-of-hazardous-material-from-piping-systems/.

Incident 6: Loss of Containment Due to the Failure to Identify Hazards (Hazard Identification)

ExxonMobil Refinery, Beaumont, Texas

A fire occurred while removing heat exchanger heads using an acetylene torch to burn bolts.

April 17, 2013; two fatalities, ten injuries

The following information is gleaned from media reports, reports from federal agencies, and court documents. This was a very tragic incident that occurred, even though a lot of good people were using what they understood to be good procedures at the time. This is intended to be a positive discussion of this incident, emphasizing the Key Lessons Learned to avoid a repeat of this type of event at other facilities around the world.

The following is extracted from a Press Release from the US Department of Justice (DOJ) and the US Environmental Protection Agency (EPA) detailing the results of the April 17, 2013, fire.

> "WASHINGTON – The US Department of Justice (DOJ) and the US Environmental Protection Agency (EPA) announced a settlement with ExxonMobil Oil Corporation (ExxonMobil) today to resolve federal Clean Air Act claims arising from a 2013 fire at the company's oil refinery in Beaumont, Texas that killed two employees and injured ten others. In a complaint filed today with the settlement, the United States alleges that the company violated Section 112(r) of the Clean Air Act, which requires measures to prevent accidental releases of extremely hazardous substances that can have serious public health and environmental consequences.
>
> The April 17, 2013, fire at the refinery occurred when workers used a torch to remove bolts from the top, or "head," of a device called a heat exchanger. The torch ignited hydrocarbons released from the head. EPA's inspection following the incident disclosed violations of Section 112(r) and of the regulations known as the Chemical Accident Prevention provisions."

The diesel hydrotreater unit where the fire occurred had been shut down for routine maintenance during a planned turnaround at the time of the accident. As one may expect, the legal actions went back and forth following this tragic fire, with ExxonMobil being sued by the injured parties or by their successors, and ExxonMobil filing suit against the contract companies for their actions, actions that ExxonMobil believed either caused the incident or contributed to the exchanger fire that resulted.

The suits revolve around the task that was performed on April 17, 2013, to open an exchanger located in a bank of 12 exchangers. Using traditional impact wrenches, the contract companies attempted to remove the large-diameter exchanger head bolts, which were seized due to corrosion. This was unsuccessful, and the contractors received permission from ExxonMobil to burn the bolts with a torch. When the exchanger was opened, liquid hydrocarbons were released and quickly ignited by the torch. As a result, the 12 contract employees from two contract companies were engulfed by the flames, resulting in two contractor fatalities and 10 others injured, at least seven were hospitalized, four were listed in critical condition with burn injuries.

The suit by ExxonMobil states "The day before the fire, Clean Harbors declared the chemical clean complete," the suit states. "It is now apparent, however, that the chemical clean was not complete. Despite ExxonMobil having hired Clean Harbors to purge all hydrocarbons from the system, an estimated 30–60 gallons were released from the channel head flange as bolts from the flange were being removed by Signature. Signature used a cutting torch – as opposed to an impact wrench – to remove the bolts, and the flame from the torch ignited the residual hydrocarbons."

The OSHA also cited ExxonMobil for failure to follow their internal procedures for line and equipment opening and safe blinding practices. The OSHA citation reads as follows:

> "29 CFR 1910.119(l)(3): Employees involved in operating a process and maintenance and contract employees whose job tasks would be affected by a change in the process were not informed of and trained in the change prior to startup of the process or affected part of the process. On April 17, 2013, or at times prior thereto, ExxonMobil deviated from its MP-203 procedure for Line and Equipment Opening and Safe Blinding Practices and did not ensure that contract maintenance workers in the Catalytic Hydrodesulfurization Unit #2 (CHD-2) at its Beaumont Refinery were trained in the changes. 29 CFR 1910.147(d)(6): Prior to starting work on machines or equipment that had been locked out or tagged out, the authorized employee did not verify that isolation and de-energization of the machine or equipment had been accomplished: On April 17, 2013 ExxonMobil did not follow its management of change process when it deviated from Maintenance Procedure 203-Procedure for Line Equipment Opening and Safe Blinding during the performance of maintenance on the E-1 heat exchangers in the Catalytic Hydrodesulphurization Unit #2 (CHD-2) of the ExxonMobil Refinery in Beaumont, Texas."

Several of these citations were contested, and this does not indicate the result of these challenges. Additional details can be found on the OSHA website. According

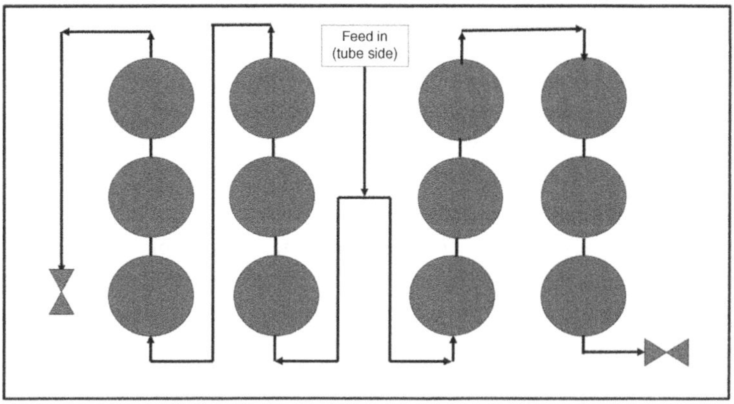

Figure 11.11 Illustration of generic hydrotreater reactor feed/product "stacked" exchangers.

to court records, two of the other companies also received OSHA citations for their role in the accident. Again, all of these can be contested, and the result of these challenges can be found online.

What we know about this incident is that the exchanger was found to be almost completely fouled, and the chemical cleaning was not successful. The plugged exchanger trapped liquid hydrocarbons in the exchanger, which were released when the bolts on the exchanger shell were burned with the torch. And of course, the flame from the torch was the ignition source, resulting in this tragic fire. It has also been reported that the exchanger had been gas tested prior to the permit being issued. However, the liquid hydrocarbons were thought to be combustible with a higher flash point and were not picked up on the gas test.

This is a real tragedy; however, there are some valuable lessons to learn, and if applied in our industry, this should not happen again. This is exactly why this case study is included in this textbook – to share the lessons learned and hopefully prevent another incident like this from occurring.

Figure 11.11 illustrates the complexity and congestion when working with a bank of stacked exchangers. This image is not from an incident and is only used here to demonstrate the difficulty of clearing these exchangers. In most hydrotreater units, the extremely hot reactor bottoms stream is on the tube side of the exchangers, preheating the reactor feed on the shell side. Not only is the flow complicated, but the congestion within the structure also hampers access.

Key Lessons Learned (Author's Comments)

This is a recap of several of the Key Lessons Learned in this incident, consolidated with other learnings from OPE case studies covered in this book. Due to the tragic outcomes of these incidents and to help stress the importance of full acceptance and compliance, there will be some duplication with the key learnings from other chapters.

- Verification of equipment isolation and zero energy must be done before turning equipment over to maintenance personnel for equipment opening. This includes complete separation from operating equipment and verification that there is no trapped energy in the form of internal pressure or trapped flammable or toxic liquids or vapors.
- Isolation should normally be achieved by either double block and bleed valves or by the installation of figure-8 blinds. Blinding should always be the technique of choice since it provides full and complete isolation. If double block and bleed valves are used, the bleed valve between the upstream and downstream valve should normally be car-sealed open.

  If there is a potential of creating a "trapped blind," vented blinds should be used to ensure the trapped energy can be safely vented away.
- All vents and drains must be verified as clear and unobstructed and must be in the fully open position to demonstrate that the equipment is at a zero-energy state. This includes a survey to ensure that there is no liquid trapped in piping low areas that may be released upon pipe opening.
- Hot work should never be used for equipment opening. However, if necessary and after a risk analysis to determine other potential options, and after a successful area gas test, the technique of hot bolting may be used, provided there is no other hot work in the immediate area.

  Hot bolting is burning only one bolt at a time, and reinstalling a new bolt in its place, tightening the new bolt or torquing the bolt, if necessary, before removing the next bolt. This process should be done until all bolts have been replaced. Following this, all hot work should be discontinued, and the equipment can then be opened the traditional way, with cold work. The advantage of using this method is that the seal is never broken while hot work is being used.
- Only fully qualified gas testers and properly calibrated and bump-tested gas test equipment should be used to conduct gas testing. The gas test should always test for oxygen first since most gas test machines require oxygen to be present before they can provide an accurate reading for flammables (% LEL).
- Always target for a completed scaffold that is properly designed for the work to be done and fully meets the OSHA requirements for a scaffold, including both primary and secondary egress. If possible, the scaffold should be constructed so that independent fall protection is not required (i.e., what is referred to as a "green tag" scaffold).
- Workers on the scaffolds and in the immediate area should be minimized to only those involved in the equipment opening during the opening of flanges, access plates, or heat exchanger heads.

- When gas testing equipment is in preparation for opening, be aware of the potential of trapped combustible liquids. Trapped combustible liquids at ambient temperature (liquids with a flash point of more than 100 °F) (38 °C), such as diesel or process gas oil, will not be accurately read on a flammable gas test (gas test for % LEL).

  A gas test above combustible liquids at ambient temperature will read 0 LEL, even though there may be a large quantity of combustible liquid in the equipment. However, if this equipment is heated to a temperature above the liquid's flash point, possibly due to an ambient temperature increase, or due to heating of any kind, the liquid will vaporize and generate flammable vapors. If an ignition source is nearby, the vapors can ignite readily and can result in a major fire or explosion.

Additional References

Beaumont Enterprise, 3 Companies Given Citations, Fines for Fatal Flash Fire at Refinery, https://www.beaumontenterprise.com/news/article/3-companies-given-citations-fines-for-fatal-4993056.php.

Environment News Service, International News Service, ExxonMobil Fined $616,000 for Fire that Killed Two, https://ens-newswire.com/exxonmobil-fined-616000-for-fire-that-killed-two/.

In the United States District Court for the Eastern District of Texas, Beaumont Division, Complaint: United States of America, Plaintiff, v. ExxonMobil Oil Corporation, Defendant, https://www.justice.gov/enrd/consent-decree/file/1141321/dl.

In the United States District Court for the Eastern District of Texas, Beaumont Division, Consent Decree: United States of America, Plaintiff, v. ExxonMobil Oil Corporation, Defendant, https://www.justice.gov/enrd/consent-decree/file/1141316/dl.

Law Offices of Gregory Jordan, ExxonMobil Sues Industrial Services Companies for Breach of Contract Over Refinery Fire, https://www.theaustintriallawyer.com/2015/01/exxonmobil-sues-industrial-services-companies-for-breach-of-contract-over-refinery-fire/.

Southeast Texas Record – Thursday, January 23, 2025, ExxonMobil Says Fire Caused by Cleaning Companies, https://setexasrecord.com/stories/510624622-exxonmobil-says-fire-caused-by-cleaning-companies.

US Department of Justice (DOJ) and the US Environmental Protection Agency (EPA), ExxonMobil to Pay Civil Penalty and Take Remedial Measures to Resolve Clean Air Act Violations Stemming from Deadly Fire at Texas Refinery, https://www.epa.gov/newsreleases/exxonmobil-pay-civil-penalty-and-take-remedial-measures-resolve-clean-air-act.

US Occupational Safety and Health Administration (OSHA), Violation Detail: Standard Cited: 19100119 F04 Process Safety Management of Highly Hazardous Chemicals, https://www.osha.gov/ords/imis/establishment.violation_detail?id=901522.015&citation_id=01001.

Incident 7: Loss of Containment Due to the Failure to Identify Hazards (Hazard Identification)

ConAgra Slim Jim production facility, Garner, North Carolina
June 9, 2009; four fatalities, three critical burn injuries, and several other injuries (sixty-seven were sent to the hospital)

A natural gas explosion occurred at this facility due to the accumulation of natural gas during a "gas blow" of a new natural gas supply pipeline.

The CSB investigated this incident. The following is a summary of this final report. Excerpts from the final report are also included below. A link to the CSB report is included in the "Additional References" section of this chapter.

> Note: The explosion at Conagra on June 9, 2009, which killed four people, was followed by a very similar event that occurred at the Kleen Energy Power Plant in Middletown, Connecticut, on February 7, 2010. In the Kleen Energy incident, another six people were killed, and dozens more were injured. As a result of these two incidents, the US Chemical Safety and Hazard Investigation Board (CSB) issued an Urgent Recommendation. See the "Additional References" section below for a link to this report.

On June 9, 2009, a natural gas explosion at the ConAgra Slim Jim production facility in Garner, North Carolina, caused four deaths, three critical life-threatening burn injuries, and other injuries that sent a total of 67 people to the hospital. In preliminary findings, CSB investigators determined that the catastrophic explosion resulted from the accumulation of significant amounts of natural gas that had been purged indoors from a new 120-foot-long pipe during the startup of a new water heater in the plant that made Slim Jims, a popular beef jerky product.

During pipe purging, workers feed pressurized gas into a pipe to displace air or other gases so that only pure fuel gas remains in the piping when connected to an appliance such as a water heater or boiler.

The accident occurred during the installation of a new fuel gas-fired industrial water heater in the plant's interior utility room. Five days before the accident, a new section of 3-inch steel piping to provide natural gas to the heater was tied into a 6-inch natural gas supply line located on the roof. The new natural gas piping ran horizontally over 120 feet along the roof and descended into the utility room.

On the day of the accident, a worker from Energy Systems Analysts (ESA), the water heater manufacturer, attempted to purge the new gas line by using natural gas to displace the air. This was done by removing threaded fittings and creating one or more pipe openings near the heater. The worker then opened a quarter-turn valve to control the release of purged gases. ESA reported that it was the company's normal practice to purge fuel gas piping directly into the room or area when installing gas-fired equipment. Code officials and other parties told the CSB they believe this practice is common.

The purged fuel gas was vented indoors into the utility room, which was ventilated by an exhaust fan. However, no assessment was made of the adequacy of the ventilation in comparison to the rate of the gas release; whether a dangerous accumulation of flammable gas had occurred could have been most accurately verified by taking direct measurements inside the utility room using a combustible gas detector. Because of the difficulties in lighting the water heater, personnel perceived that the gas line was not effectively purged of air. Therefore, purging was conducted intermittently for up to two-and-a-half hours. Refer to Figure 11.12 for a photo of the new gas-fired water heater and gas piping.

CSB Chairman John Bresland said, "The board is very concerned that companies across the country continue to purge pipes indoors, and this evening, we will consider recommendations to the NFPA, the American Gas Association (AGA), and the International Code Council (ICC). Currently, the codes of the NFPA and ICC do not require gases to be vented outdoors or define adequate ventilation or hazardous conditions, nor do they require the use of combustible gas

US Chemical Safety and Hazard Investigation Board (CSB). Public Domain.

Figure 11.12 Photograph of the new gas-fired water heater and piping, indicating points where gas was likely released into the building.

US Chemical Safety and Hazard Investigation Board (CSB). Public Domain.

Figure 11.13 ConAgra Slim Jim production facility in Garner, North Carolina, after a devastating natural gas explosion that occurred during the purge of a new natural gas supply pipeline.

detectors during these operations. The CSB recommendations, if adopted, would urge that these things be done."

CSB investigations supervisor Donald Holmstrom said his team made recommendations to the board during the course of the ConAgra investigation after discovering gaps in the fuel gas codes. "Purging flammable gases into building interiors is a recipe for disaster. At ConAgra, we determined the accident would not have happened had the gas been vented safely outdoors through a hose or pipe." Mr. Holmstrom noted that since the June 2009 accident, "ConAgra has instituted strict policies on purging, requiring it to be done at safe outdoor locations." Source http://www.csb.gov/.

Refer to Figure 11.13 for a photograph of the ConAgra facility after the explosion.

Key Lessons from This Incident (as Reported by the CSB)

1. Purging new or existing gas piping into buildings can be highly hazardous due to the possible accumulation of gas above the lower explosive limit (LEL) and the associated danger of fire and explosion. Wherever practicable, directly vent the purged gases to a safe location outdoors, away from people and ignition sources. This can be done using a temporary hose or piping or permanently installed vent pipes, depending on the facility design.
2. Purging indoors should only be done in limited circumstances where purging outdoors is not practicable. In such cases, nonessential personnel should be

evacuated; all ignition sources should be controlled or eliminated; and ventilation should be adequate to maintain the gas concentration well below the LEL at all times.

3. Never rely on odor alone to detect releases of fuel gases. An odorant is typically added to fuel gases, such as natural gas and propane, to warn workers and consumers of releases. However, the perception of odor is highly subjective and varies from one person to another. People also become desensitized to odor during prolonged exposures. Additionally, new gas pipes and containers can react with or otherwise remove the odorant, an effect known as "odor fade."
4. Always use combustible gas detectors to monitor the gas concentration during purging operations. To provide the most accurate information about combustible gas levels, sampling should be conducted frequently or continuously at appropriate locations.
5. Ensure personnel involved in gas purging operations are fully trained and knowledgeable about safe gas venting practices, the proper use of gas detectors, and the danger of relying on the sense of smell alone to detect gas releases. Include training on the problem of odor fade in new gas piping systems.

Author's Comments

If possible, try to avoid gas blows. If this process is allowed, ensure the safe route of the gas.

If it is permitted to be carried out, gas blows before startup should never be directed toward enclosed areas or places where people are present. It should be vented high above structures and enclosures to allow the gas to safely disperse into the atmosphere. This is simply common sense, and it ensures that the gas is released well above all surrounding equipment, where no people or equipment will be involved if an ignition occurs.

Additionally, all ignition sources should be removed from the area before starting the operation and kept away until well after it is completed.

Additional References

Bakery & Snacks, ConAgra to Close Blast Damaged Slim Jim Plant, https://www.bakeryandsnacks.com/Article/2010/03/05/ConAgra-to-close-blast-damaged-Slim-Jim-plant/.

Firehouse, Explosion and Collapse at North Carolina Food-Processing Plant, https://www.firehouse.com/home/article/10470653/t-approximately-1127-am-on-tuesday-june-9-2009-an-explosion-ripped-through-the-425000-square-foot-conagra-foods-inc-processing-plant-in-garner-nc-the-facility-is-the-sole-source-of-slim-jim-snack-foods-for-the-company-product-line-the-initial-b.

US Chemical Safety and Hazard Investigation Board (CSB), ConAgra Natural Gas Explosion and Ammonia Release, https://www.csb.gov/conagra-natural-gas-explosion-and-ammonia-release/.
US Chemical Safety and Hazard Investigation Board (CSB), Kleen Energy Natural Gas Explosion (Includes CSB 'Urgent Recommendation'), https://www.csb.gov/kleen-energy-natural-gas-explosion/.

Incident 8: Loss of Containment Due to the Failure to Identify Hazards (Hazard Identification)

Huntsman Chemical, Port Neches, Jan 11, 2004

Thermal Decomposition of trapped liquid resulted in a fire while the piping was being prepared for maintenance work.

January 11, 2004; this incident resulted in eight injuries, including two serious burn injuries

While preparing a piping system for maintenance, the line was drained and steamed in preparation for opening. A low spot in the line accumulated liquid hydrocarbons, which expanded when heated and caused the pipe to rupture. The resulting vapor cloud ignited. The CSB investigated this incident and issued a Safety Bulletin to share its findings.

Huntsman Chemical Company, located in Port Neches, Texas, experienced an explosion and flash fire on January 11, 2004. In this incident, eight workers were clearing a 6-inch pipeline of residual chemicals by purging with nitrogen, followed by high-pressure steam. The pipeline was part of a unit that produces methyl tert-butyl ether (MTBE), a fuel additive. The fire burned two plant workers and caused minor injuries to six others. Four workers, including the two who suffered second-degree burns, were treated and released at local hospitals. It was also reported that the men wore Nomex coveralls, preventing more serious injuries.

The CSB investigators reported that the fire occurred as workers prepared the pipeline for maintenance. They noted that the piping design included a long, shallow dip where residual and thermally reactive feedstock chemicals had collected. The CSB stated that the mixture was trapped in a low point along a 900-foot-long, 6-inch process pipe.

The piping had accumulated a residual chemical that was thermally reactive, which, unknown to plant personnel, reacted when they steamed the piping to clear the line of residual hydrocarbon. Plant personnel mistakenly believed that a nitrogen gas purge removed all liquid from the pipe before the steam purge. However, about 350 gallons of a thermally reactive peroxide/alcohol mixture remained in a previously unidentified low-point area of the piping. As the steam heated the trapped peroxide, it decomposed exothermically and overpressurized the pipe, causing it to rupture. The venting of flammable vapor ignited, leading to a flash fire that burned employees and damaged nearby equipment. The fire was suppressed by the automatic deluge system and the fire brigade.

As it turned out, the piping involved in the accident was rarely used at the ten-year-old facility. When it was last used in September, it was purged with nitrogen gas; however, not all the chemicals had been removed. Huntsman also reported that they have investigated similar plant conditions like this one and have implemented corrective actions to help prevent a recurrence elsewhere.

The pipe routing included a low-point segment nearly 300 feet long, which held about 350 gallons of the reactive chemical and was partially obscured by an overhead pipe rack. The existing procedure called for an extra nitrogen purge as a precaution, followed by a steam purge to remove any remaining hydrocarbon vapor. The operators were unaware that the nitrogen gas purging, required by the operating procedure, was not effective in removing all the peroxide and alcohol mixture in the low-point segment of the piping.

The CSB identified low point drains in the low point area of the piping, indicating that the liquid could have been safely removed without performing a steam purge. This was confirmed by reviewing the as-built drawings and conducting a walkdown of the entire peroxide/alcohol transfer pipe section.

During the use of the generic steam purging procedure, the line temperature exceeded 380 °F, which is well above the peroxide's decomposition temperature. The high-temperature steam circulated through the piping system for nearly 1.5 hours, causing the trapped peroxide to start decomposing. This pressure buildup in the piping caused a valve gasket to fail, releasing flammable vapor that led to a fire. Around the same time, the pipe upstream of the leaking valve ruptured, releasing additional flammable liquid and vapor, further fueling the fire. The explosion and overpressure that followed caused additional damage to equipment in the area.

Complex pipe routing and equipment configurations often contain low points that trap hazardous material. Flushing the pipe with water or other suitable liquid may effectively remove the material but generate a significant quantity of waste liquid. Furthermore, liquid that remains in low points or equipment may cause unsafe conditions or degrade piping. To reduce waste volume, Huntsman revised the procedures to substitute inert gas purging for water flushing. However, as in this case, purging with inert gas or steam may not remove trapped liquid.

The revised procedure failed to address the importance of identifying low points in the piping and failed to require the use of low point drains to remove trapped hazardous liquids. Proper identification and use of low-point drains ensure the complete removal of hazardous liquids. The revised inert gas purging procedure was applied to process pipes containing thermally unstable liquids without adequately considering the hazards if these liquids remained in the pipe during the steam purge. High-temperature steam, when used to remove traces of flammable vapor from the pipe before performing hot work, may cause thermally unstable liquids to decompose violently. This is precisely what occurred with the peroxide/alcohol mixture in this incident.

Key Lessons Learned

The nonroutine opening of chemical process piping and equipment can be very dangerous and should never be treated as regular work. Personnel at facilities that handle hazardous chemicals should conduct a thorough walkdown of all piping and components between the isolation devices. As-built drawings should be updated as needed.

Use the as-built drawings of the affected piping to identify all branch connections, isolation valves, low-point drains, and high-point vents. A specific written procedure for removing, flushing, and purging hazardous material from the system should be developed. If liquid stays in the system, consider the consequences of the remaining liquid after finishing the work.

Unit-specific procedures should be established and employed to guarantee the safe execution of nonroutine activities, such as steam purging of process lines that handle thermally unstable liquids.

Facilities should:

- Review planned steps against unit-specific hazards.
- Update unit-specific procedures as necessary to address unique characteristics of the activity, especially when modifications are involved.
- If a modification activity includes the use of generic safety procedures, clearly identify applicable constraints (e.g., specify additional inspection steps to verify removal of all energy sources, including thermally reactive chemicals)

Author's Comments

I was surprised to hear the CSB report that the operators were unaware that residual chemicals used in their process could react violently at high temperatures, such as when steaming the equipment with high-pressure steam. This information should be a core part of the operators' training and certification process. It should also be included on the Safety Data Sheet (SDS) and in their operating procedures, including critical steps to avoid exposure to these high temperatures.

The CSB's recommendation to use the unit drawings to locate and provide both low-point drains and high-point vents is spot on. These are useful drawings, but they require a lot of effort to maintain them current. They become extremely helpful when maintained properly, especially in congested pipe racks or other busy equipment areas.

Additional References

Environmental and Natural Resources Division – US Department of Justice, TPC Group LLC Investigation-Port Neches Explosion, https://www.justice.gov/enrd/case/tpc-group-llc-investigation-port-neches-explosion.

The Texas Tribune, Federal Report Identifies Cause of 2019 Port Neches Chemical Plant Explosion, https://www.texastribune.org/2022/12/20/texas-chemical-plant-explosion-tpc-port-neches/.

US Chemical Safety and Hazard Investigation Board (CSB), The Danger of Popcorn Polymer: Incident at the TPC Group Chemical Plant, https://www.csb.gov/videos/?SID=3607.

US Chemical Safety and Hazard Investigation Board (CSB), TPC Port Neches Explosions and Fire, https://www.csb.gov/tpc-port-neches-explosions-and-fire/.

Incident 9: Loss of Containment Due to the Failure to Identify Hazards (Hazard Identification)

ExxonMobil Terminal, Nagoya, Japan

A gasoline Tank Fire occurred in a tank being cleaned while hot work was in progress in an adjacent gasoline tank.

August 29, 2003; six fatalities and one injury requiring hospitalization

This incident occurred while a gasoline storage tank was undergoing hot work to convert the roof design, while an adjacent storage tank was being cleaned for opening.

This case study is based on a paper presented by: KOBAYASHI, Mitsuo (Office K) & TAMURA, Masamitsu (Center for Risk Management and Safety Sciences, Yokohama National University). Other quoted data sources include the following:

- Accident prevention technical center, urgent report; Outline of factory fires that have occurred frequently. Safety & Tomorrow. No. 92. pp. 17–27 (2003).
- Large-scale fire in August–September 2003, Fire. No. 266, pp. 5–7(2003).
- Chunichi Newspaper (2003).

On August 29, 2003, work was in progress to convert two floating roof gasoline storage tanks to internal floating roof tanks. Tank 2 was the larger of the two, with a capacity of about 31,500 bbl, while Tank 24 had a capacity of approximately 13,800 bbl. The two gasoline tanks were enclosed within a high firewall surrounding both tanks. A smaller 12-inch-high curb was installed inside the firewall between the tanks. There was a single explosion-proof gas detector inside the firewall of Tank 2, with a non-explosion-proof buzzer near the manway access of Tank 24.

Tank 24 had been emptied for about a month, had undergone gas testing, and was confirmed gas-free. Work permits had been issued, allowing workers to enter Tank 24 and begin internal hot work. Five workers were inside, performing hot work to make the necessary internal modifications to convert the tank into an internal floating roof tank.

While hot work was ongoing on Tank 24, work began on Tank 2 to finish transfer the remaining gasoline using a pneumatically operated portable pump. The gasoline was being moved to another storage tank with a tank truck. One of the manway plates had been removed from Tank 2 the previous afternoon and reinstalled overnight. It was opened again on the morning of the 29th, and workers continued transferring gasoline from the tank bottom with the portable pump. A hose had been inserted into the open manway of Tank 2, and gasoline extraction was in progress at the time of the incident.

As Tank 2 was nearly empty, three workers entered the tank to rinse the tank bottom with water hoses to push the remaining gasoline toward the portable pump. The pump extracted the mixture of waste oil and gasoline. In mid-afternoon, with three workers inside Tank 2 and five inside Tank 24, another contractor removed one of the roof manways on Tank 2 and then prepared to remove the second roof manway. Suddenly, opening the roof manway increased airflow into the tank through the roof and boosted the release of gasoline vapors from Tank 2.

A few minutes later, a contractor exited Tank 24, heard the gas detector alert, and warned others to exit Tank 24. At about this time, five Tank 24 workers outside the tank witnessed a gasoline haze approximately three feet high. Six of the Tank 24 workers began to evacuate the area. Witnesses saw the ignition start near the Tank 24 area. However, the fire quickly engulfed both tanks.

Six workers at both tanks were killed, three in Tank 2 and three in Tank 24. One person was injured, requiring hospitalization. It is believed the gasoline vapors were ignited by the hot work that was in progress in Tank 24, or by the non-explosion-proof buzzer, or possibly by someone disconnecting the power cable from the transformer. One gas detector sounded, but it was too late. See Figure 11.14 for a sketch of where this tragic incident occurred.

The following is a direct quote from the Japanese Failure Knowledge Database report:

> "The cause was clearly a management system error. Draw-off work for light hydrocarbons at an adjacent tank was carried out with the manhole open. At the same time, hot work was carried out with a gas detector set up near an adjacent tank. There is a lack of risk management based on the prediction of the generation of gasoline vapor from gasoline draw-off work at an adjacent tank, or the generation of combustible gas from venting gas from another tank and drainage work, etc. There might have been a lack

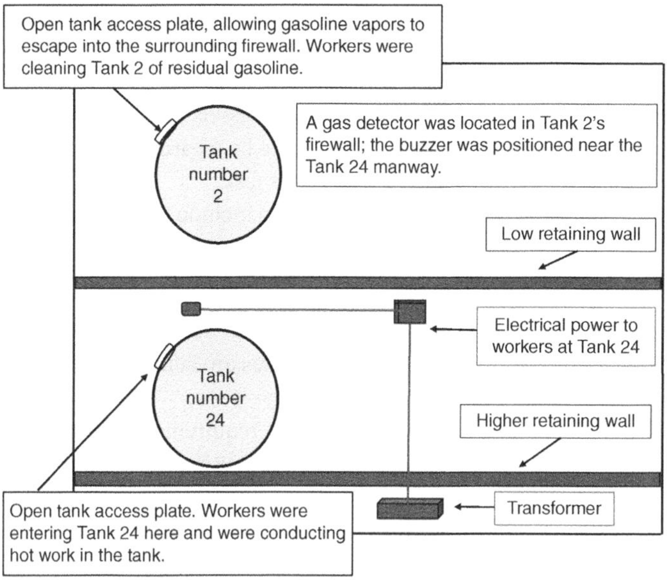

Figure 11.14 Area near Nagoya, Japan, where this tank fire occurred. This photograph is courtesy of the Japanese Failure Knowledge Database.

of consciousness of the hazard and an insufficient grasp of the details of construction.

1. It is natural when light hydrocarbons such as gasoline are released into the atmosphere, combustible vapor is generated. This vapor spreads. At the opposite side of the manhole of the tank, the vapor also spreads. This is a basic point to be understood in handling a liquid with a low ignition point, such as gasoline.
2. Safety countermeasures for the construction of hazardous facilities are primarily the responsibility of the owner. Even if the subcontractor is knowledgeable, the owner must have superior knowledge of hazardous substances and the situation in the facilities.

Fundamentally, details of construction are grasped every day, and it is necessary to consider a construction method and a schedule that are suitable. In the case of this accident, it should be necessary to stop hot work in the adjacent area as gasoline was released from the tank."

Author's Comments

I fully support the Japanese Failure Knowledge Database, as this was a management system failure at the time of the incident. I am proud to say that I was an employee of ExxonMobil during this period. Although I worked at a different site, I witnessed firsthand the changes the company made afterward. The company developed and implemented an industry-leading practice, also known as a Management System Standard, for safely removing storage tanks from service, clearing their contents, and opening these large process vessels to the atmosphere. This best practice was recognized then and continues to be regarded as one of the industry's top standards for performing this hazardous task.

Just a few examples of some of the new requirements include the following:

Note: These are summarized examples only; this is not intended to be a complete list.

- The requirement that a Job Supervisor will be present during all the critical steps in the procedure to ensure that the procedures are being implemented correctly
- A pre-cleaning conference to ensure that all safety requirements are in place
- Verification that the tank has been emptied BEFORE any manways are removed
- All ignition sources are removed or made safe before the tank is opened
- Residual products are removed from the tank WITHOUT personnel entering the tank
- Ensuring proper bonding and grounding to prevent ignition by static electricity
- Ensuring proper ventilation of the tank to remove any remaining vapors
- Strict requirements for LEL, H_2S, and other hazardous or toxic levels before tank entry
- Specifying proper PPE for workers

In hindsight, work on Tank 24 should have been stopped, and the work permit revoked, while all tank-clearing activities on Tank 2 were ongoing. Due to the proximity of these two tanks, one to another and within a common tank firewall or berm, there should have been clear communication between the two work teams. This includes ensuring that all hot work and other spark-producing activities related to both tanks are ceased until both tanks are fully isolated, properly gas tested, and confirmed to be hydrocarbon-free. Of course, as the saying goes, hindsight is always 20/20. Therefore, strict procedures and safety standards must be in place to guide these decisions and actions to help prevent another tragedy like this.

Additional References

Japanese Failure Knowledge Database – Case Details, https://www.shippai.org/fkd/en/cfen/CC1300011.html.

Unified Facilities Guide Specifications (USGS), Cleaning of Petroleum Storage Tanks, https://www.wbdg.org/FFC/DOD/UFGS/UFGS%2033%2001%2050.55.pdf.

UPI News, Exxon Mobil Plant Blast Kills Four Workers, https://www.upi.com/Archives/2003/08/29/Exxon-Mobil-plant-blast-kills-four-workers/3041062129600/#:~:text=NAGOYA%2C%20Japan%2C%20Aug.,facility%20of%20Exxon%20Mobil%20Corp.
US Occupational Safety and Health Administration (OSHA), Major Work Activities for Tank Cleaning Operations, https://www.osha.gov/storage-tanks/work-activities.

Series II. Loss of Containment Due to the Failure to Properly Identify Equipment for Opening or the Workers Cutting or Opening the Wrong Equipment

Incident 10: Loss of Containment Due to the Failure to Properly Identify the Equipment for Opening, or Workers Cutting or Opening the Wrong Equipment

Martinez Refining Company, Martinez, California

A release and a significant fire occurred while opening the process equipment for maintenance.

February 1, 2025, Local media reported that six workers were reported injured, four of whom were treated at a local hospital and released (note this is preliminary since the incident just recently occurred).

On Saturday, February 1, 2025, a hydrocarbon leak at the Martinez Refining Company caused a fire that resulted in six minor injuries, four of which were treated at a local hospital before being released. Local news reported that PBF purchased the refinery in 2020, marking this as the third significant incident since the acquisition. The media also reported that there have been additional spills and releases following the purchase.

The large fire released significant plumes of smoke into the local community, leading to a shelter-in-place order and a public health advisory. Contra Costa asked the local community to "Please go inside, close all windows and doors, turn off all heaters, air conditioners, and fans. If not using the fireplace, close fireplace dampers and vents, and cover cracks around doors and windows with tape or damp towels."

In a report provided by The Martinez Refining Company on February 4, 2025, it was stated that when the incident occurred, two workers were OPE ahead of the scheduled maintenance on one of their process units. The unit had been shut down several days prior to the incident, and when the equipment was opened, hydrocarbons began to leak into the atmosphere. The workers immediately evacuated before the spilled material ignited, resulting in a spreading fire. The workers were transported to the hospital for treatment. Except for the required utility support systems, the refinery has shut down.

A report dated May 6, 2025, prepared by an outside consultant and submitted to Contra Costa County Health Services, concluded that the fire at the Martinez Refinery on February 1, 2025, was caused by contractors opening the wrong flange on a pipeline while attempting to install an isolation blind to isolate the piping circuit for maintenance. As illustrated in Figure 11.15, the drawing submitted by

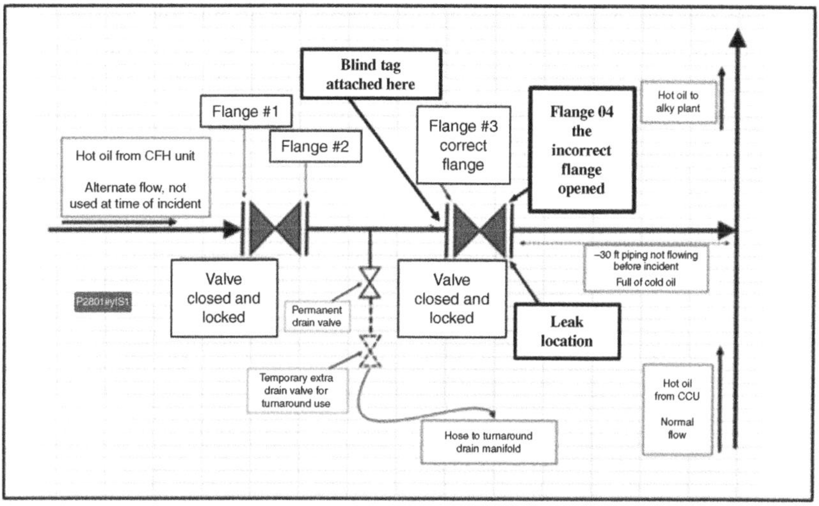

Figure 11.15 Image of the Martinez Refinery piping circuit and the flange that was inadvertently opened. Photo by Contra Costa County Health Services (California).

Contra Costa Health Services indicated that the flange to be opened was tagged with a blind tag.

The report indicated that the piping circuit had been properly isolated, vented, drained, and prepared for opening. The contractors began opening the correct flange on the upstream side of the downstream isolation valve; however, they inadvertently opened the downstream flange of the same valve instead. This side of the valve had not been prepared and was still under pressure.

While working from a scaffold, the contractors released hot oil at approximately 630 °F (332 °C), which ignited about a minute after the release, likely due to autoignition. Fortunately, the workers managed to escape before the ignition occurred. This significant fire caused extensive damage to the refinery and resulted in a prolonged shutdown for several units. See Figure 11.15 for an image of the piping circuit that was involved and the flange that was inadvertently opened.

One can see that the piping circuit was properly prepared for opening and blind insertion. The isolation valves were closed and locked in the closed position. The circuit between the isolation valves was properly drained and depressurized, and the point of blinding was properly tagged. However, the flange on the pressurized side of the downstream isolation valve was opened, releasing hot and light hydrocarbons to the atmosphere, where they quickly ignited, probably by autoignition, resulting in the major refinery fire and injuries to personnel.

The refinery fire team quickly responded to initiate firefighting efforts, and mutual aid was called, involving local industry and the Contra Costa Fire

Protection District. A Unified Command, comprising of multiple agencies and organizations, worked together effectively and was released at about 8:30 p.m. when responsibility transitioned back to the refinery teams. See Figures 11.16 and 11.17 for photographs of the February 1, 2025, fire that occurred at the Martinez Refinery in Martinez, California.

Courtesy of Contra Costa News and D. Darling.

Figure 11.16 Martinez Refining Company during major refinery fire, which reportedly occurred on February 1, 2024, while OPE for maintenance.

Courtesy of Contra Costa News and Shannon Mitchell-Frades.

Figure 11.17 Martinez Refining Company during a major refinery fire, which reportedly occurred on February 1, 2024, while OPE for maintenance.

Local media quoted Contra Costa County Supervisor Shanelle Scales-Preston during a news conference, stating, "This is the third major incident since PBF Energy purchased the refinery in 2020, and many more spills and releases have occurred since then. This is not acceptable, and I am working to get answers for our community. Currently, the County and other regulatory agencies will conduct an independent investigation into what caused the fire, what violations may have occurred, what potential environmental hazards linger, and what necessary improvements are needed to ensure public and workers' safety and health. In addition, I am requesting a Full Facility Audit under the provisions of the County's Industrial Safety Ordinance."

This incident occurred this past weekend, and so far, the Chemical Safety Board has not indicated whether it is sending a team of investigators to the site. We know that the CSB has other significant incidents under investigation and that this role is very important in our industry. We look forward to receiving those reports and learning the key lessons. The CSB selects the incidents it decides to investigate very carefully, likely due to limited staff. However, when the CSB investigates, it publishes very detailed reports that can make a difference in our industry and across all industries that handle flammable or toxic chemicals.

The State of California Contra Costa Health Services has indicated that they are investigating the Martinez Refinery incident. They also have a Public Records Access process for which I requested the incident reports when they are completed. The information above is preliminary; if additional information becomes available before this goes to the publisher, it will be included here. So far, we know that hydrocarbons were released while craftsmen were OPE, resulting in a major fire, a significant environmental release, six minor injuries, and an extended shutdown of the refinery.

Key Lessons Learned (Pending Release of the Contra Costa Investigation Report)

I realize this incident just happened this week, and we certainly don't have all the details that will emerge when a thorough investigation is completed. However, what I am attempting to do here is to highlight the frequency and seriousness of work involving the OPE, as well as the many details that can and must be in place before beginning this task. This involves proper and detailed job planning, job-site preparation, and field execution, including the following:

- Always verify complete equipment isolation from other live equipment.
- Always verify zero energy regarding trapped hydrocarbons and possible leaking valves between live equipment and offline equipment.
- Always ensure absolute identification and proper tagging of each flange and access plate to be opened.
  - Each flange should be tagged on the depressurized side of the flange to be opened.
  - This should be done AFTER the operations organization prepares and before any flange is opened.
- Always ensure effective and clear communication during the joint job-site visit, and both the permit issuer and permit receiver should review each flange and/or access plate that needs to be opened.
  - Point out the unique tag on each flange or access plate indicating that it has been approved for opening.
  - The permit acceptor should be reminded that no flanges will be opened unless this tag is attached.

- Always provide and use proper PPE appropriate for the product's characteristics.
- Always anticipate that the line may be pressurized – remind the craftsperson never to remove all bolts from a flange; instead, crack the flange first, allowing the craftsmen to quickly reclose the flange in case it is found to be pressurized.
- Avoid using hot work when opening flanges, and do not open flanges if hot work, such as burning or welding, is being conducted in the area.
- Always provide a drain pan to safely collect any small amounts of residual material for each flange being opened. Ensure the drain pan is conductive and bonded, and grounded to protect against static electricity hazards.
- Also, for products with high hydrogen sulfide concentration, always be aware of the potential for pyrophoric materials such as iron sulfide scale. These can ignite, resulting in an internal fire or smoldering, and can seriously damage the vessel or internals.
  - In these cases, ensure that a charged fire nozzle is at each manway being opened and periodically spray water into the vessel to keep the materials from drying.
  - Use caution if the vessel or internals are stainless steel. This may require untreated well water or steam condensate to prevent creating stress corrosion cracking due to chlorides in fire water.
- The operations group should periodically revisit the job in progress to verify that all safety and environmental precautions and good equipment opening practices are properly adhered to.

Additional References

Contra Costa County Health Services, Draft Final Report, Independent Incident Investigation/Root Cause Analysis of the Martinez Refining Company LLC, February 1, 2025, Loss of Process Containment and Fire, https://www.cchealth.org/home/showpublisheddocument/31810/638839587286590004.

Contra Costa Herald, Martinez Refining Company Issues Update on Refinery Fire, https://contracostaherald.com/martinez-refining-company-issues-update-on-refinery-fire/.

Contra Costa News, February 4, 2025, Contra Costa Fire Releases Incident Report from Martinez Refinery Fire, https://contracosta.news/2025/02/04/contra-costa-fire-releases-incident-report-from-martinez-refinery-fire/.

MSN News, Martinez Refinery Fire Leaves Health and Psychological Impacts Days Later, https://www.msn.com/en-us/news/us/martinez-refinery-fire-leaves-health-and-psychological-impacts-days-later/ar-AA1ymw27.

Incident 11: Loss of Containment Due to the Failure to Properly Identify the Equipment for Opening, or Workers Cutting or Opening the Wrong Equipment

PEMEX Refinery, Deer Park, Texas

Hydrogen sulfide was released while removing blinds.

October 10, 2024; two fatalities, thirteen were injured, and the cities of Deer Park and Pasadena were ordered to shelter in place

I have briefly mentioned this incident a couple of times thus far; however, this is a very recent incident involving equipment preparation and blinding to prepare a process pipeline for maintenance work. This incident occurred on October 10, 2024, and resulted in a catastrophic loss of containment of hydrogen sulfide. Two workers were killed almost instantly from the release, and thirteen additional injuries occurred. The release resulted in a shelter-in-place order for the cities of Deer Park and Pasadena for several hours until the release was abated.

The CSB sent a team to the site; an investigation is underway at the time of this writing. There is little doubt that we will soon have a detailed report on this incident and recommendations for PEMEX and the industry designed to help ensure that an incident like this one will not be repeated. The CSB released the following update on November 20, 2024:

> "During a maintenance activity at 4:23 p.m., approximately 27,000 pounds of toxic hydrogen sulfide gas were released. In addition to the two fatalities and other serious injuries, the release resulted in shelter-in-place orders issued by officials in the nearby cities of Deer Park and Pasadena, Texas, lasting for several hours, and a portion of Texas State Highway 225 was closed temporarily."
>
> CSB Chairperson Steve Owens said, "Hydrogen sulfide is a highly toxic substance that can result in serious injury and death even at low concentrations. This dangerous incident resulted in the death of two workers and put other workers and the surrounding communities at very serious risk."
>
> The CSB's update states that on the day of the incident, two contract workers at the refinery partially opened a flanged connection on piping containing hydrogen sulfide gas, which caused the release. One of these workers was fatally injured. According to the update, the work was supposed to be done on a different, isolated piping segment located about five feet away from the flange that was opened. Two other contract workers from a separate company, who were working on equipment less than 250 feet away and downwind from the release, were also affected, with one fatally injured from the hydrogen sulfide exposure.

See Figure 11.18 for a photograph of the pipe rack where the wrong flange was opened.

The immediately dangerous to life or health (IDLH) value for hydrogen sulfide is 100 ppm (parts per million). CSB notes that concentrations exceeding 1,000 ppm can cause death almost immediately. The pipeline, which was opened, the adjacent ARU7 piping segment, contained approximately 90% hydrogen sulfide gas at a gauge pressure of approximately 15 pounds per square inch. 90% H_2S expressed in ppm equals 900,000 ppm. See Figure 11.19 for an image of the PEMEX incident site showing where the H_2S flange was opened, and the location of the work crew downwind of the release, where the second person was killed.

Author's Note:
There appear to be no markings or other indications of where the new blinds should be installed. This can easily lead the workers to get onto the wrong flange and open a flange that has not been properly prepared for opening.

US Chemical Safety and Hazard Investigation Board (CSB). Public Domain.

Figure 11.18 Photograph of the PEMEX incident site and the apparent mix-up of the flanges to be opened for blind insertion. The flange, which was in service and pressurized and contained a high concentration of hydrogen sulfide gas (H_2S at 900,000 ppm), is shown on the left, while the flange, which had been depressurized and prepared for opening, is shown on the right. Unfortunately, the workers accidentally opened the flange on the left, causing a large release of hydrogen sulfide into the refinery and nearby communities. Note that there is no visible identification of which flange should be opened in this photograph. No visible tags, paint marks, or other identifications to indicate which flange should be opened. Also, the pipeline in the high-concentration hydrogen sulfide service does not seem to be identified visually. Had this highly visible identification been present, these markings might have warned the craftsmen that they were opening the wrong flange.

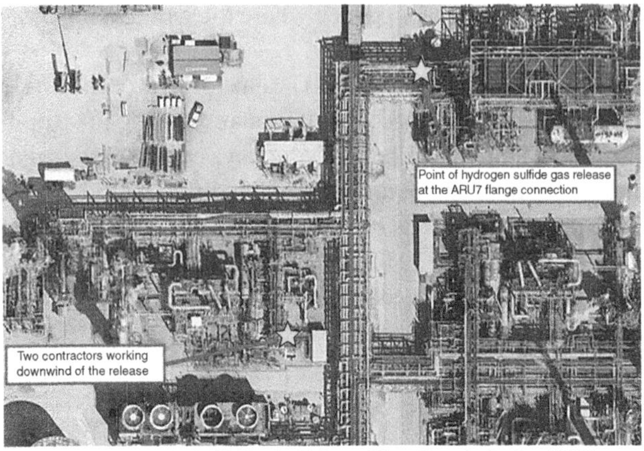

Google Maps, annotated by the US Chemical Safety and Hazard Investigation Board (CSB).
This image is from the CSB update on the PEMEX Incident investigation dated November 2024.

Figure 11.19 Photograph of the PEMEX incident site showing where the H_2S flange was opened, and the location of the work crew downwind of the release, where the second person was killed. This is an overhead view of the location of the PEMEX hydrogen sulfide release and adjacent units, with stars indicating the approximate location of the two fatally injured contract employees.

See Figure 11.19 for an image showing where the flange was opened and the location of the workers downwind of the release. One of the workers who opened the flange and one other worker who was downwind were killed in this incident.

The CSB's investigation into the incident is focused on several key areas, including:

- Permit-to-work practices
- Energy isolation procedures
- Contractor management systems
- Emergency preparedness, communication, and response systems
- Hazard analyses and risk assessments
- Maintenance procedures
- Training programs
- Respiratory protection
- Remote isolation capability

The CSB's investigation is ongoing, and the CSB will provide complete findings, analyses, and recommendations in its final investigation report."

Key Lessons Learned (Preliminary, Pending Release of the CSB Report)

I am certainly not trying to get ahead of the CSB on this one, and I do look forward to their final report and a longer list of lessons learned and recommendations to help prevent a recurrence. However, based on the update, there are some things that do stand out, and I'll focus only on these here.

It is obvious that the contract workers, while removing blinds, somehow got off onto a separate pipeline and opened a flange on a live pipeline. Unfortunately, the line that they opened appears to be associated with the Amine regeneration unit, which processed hydrogen sulfide in extremely high concentrations, typically as high as 70–80% or even higher (this equates to about 700,000–800,000 ppm or even higher). If this is true, exposure to this concentration certainly means instantaneous death to anyone exposed to these concentrations, unless they were wearing a breathing-air respirator.

This super high concentration is in perspective when it is considered that the hydrogen sulfide OSHA eight-hour personnel exposure limit (PEL) is 20 ppm, the NIOSH eight-hour exposure limit is 10 ppm, and the NIOSH IDLH is only 100 ppm.

A good practice is to identify all flanges or access plates to be opened with a unique tag developed solely for this purpose. The tag should be placed on the flange or plate to be opened after the operations group has prepared the system for opening and has verified zero energy and no obstructed vents or drains. The craftsmen should be trained to open only those flanges or access plates identified in this way.

A good practice for pipelines operating with these super concentrations of hydrogen sulfide or other highly toxic chemicals, such as hydrofluoric acid (HF), is to have the pipelines clearly identified as the pipeline travels through the piperack. This is to help ensure that anyone working near these pipelines is aware that they are there and the hazardous nature of the lines. The idea is that all workers working near these lines should recognize the extreme hazard. For example, any heavy lifts around these lines would require special planning and precautions to avoid any entanglement that could result in a release of H_2S, HF, or other highly toxic materials. Generally, this line identification is done by a uniquely colored paint stripe on the high H_2S pipelines, or with signs, labeling, or other similar identification. See Figure 11.20 for a photograph of a congested pipe rack illustrating why it is a good idea to label or identify pipelines carrying highly toxic chemicals.

Again, I am impressed by the stated scope of the ongoing investigation and look forward to the final report and recommendations.

Also, and this is very important, the workers, including contractors, should be made aware that all bolts should never be removed from a flange that is being opened. Instead, the bolts should be slowly opened, and the flanges only cracked open to verify that there is no pressure or liquid present before removing all bolts. This allows the workers to quickly reclose the flange in the event that pressure or

Courtesy of Occupational Safety and Health Administration (OSHA).

Figure 11.20 A photograph of a typical petroleum refinery or petrochemical plant pipe rack, illustrating the congestion and proximity of one piping circuit to others. This is why piping circuits carrying a high concentration of hydrogen sulfide (H_2S), hydrofluoric acid (HF), chlorine (Cl_2), or other highly toxic chemicals should be well labeled and marked with distinctive paint striping or other ways to indicate the deadly nature of the contents.

liquid is found. Another good practice is to remind the workers that they should always crack open the flange away from themselves or where others are present.

Additional References

AirGas, Safety Data Sheet – Hydrogen Sulfide, https://www.airgas.com/msds/001029.pdf.

American Industrial Hygiene Association (AIHA), CSB Releases Update on Investigation of Fatal Hydrogen Sulfide Release at Texas Refinery, https://www.aiha.org/news/241205-csb-releases-update-on-investigation-of-fatal-hydrogen-sulfide-release-at-texas-refinery.

Klinger, Connecting Flanges: 10 Common Mistakes and How to Avoid Them, Learn the Correct Methods for Making Flange Connections and Avoid Common Mistakes with Bolts and Gaskets, https://www.klinger-international.com/en/news/connecting-flanges-10-common-mistakes-and-how-to-avoid-them.

US Chemical Safety and Hazard Investigation Board (CSB), US Chemical Safety Board Releases Investigation Update into Fatal Hydrogen Sulfide Release at PEMEX Deer Park Refinery in Deer Park, Texas, https://www.csb.gov/us-chemical-safety-board-releases-investigation-update-into-fatal-hydrogen-sulfide-release-at-pemex-deer-park-refinery-in-deer-park-texas/.

US Occupational Safety and Health Administration (OSHA), Effects of Hydrogen Sulfide H_2S, https://www.osha.gov/sites/default/files/2018-12/fy15_sh-27664-sh5_Confined_Space_Handout_Effects_of_H2S.pdf.

US Occupational Safety and Health Administration (OSHA), OSHA Occupational Chemical Database – Hydrogen Sulfide, https://www.osha.gov/chemicaldata/652.

Incident 12: Loss of Containment Due to the Failure to Properly Identify the Equipment for Opening, or Workers Cutting or Opening the Wrong Equipment

BASF Chemical Plant, Ludwigshafen, Germany

October 17, 2016; four fatalities, seven serious injuries, twenty-two others injured

BASF issued a public Fire Update on December 2, 2016. The following report is based on this update and contains excerpts from the BASF report. Also, information as reported by the local District Attorney and local media is included. See the "Additional References" section at the end of this chapter, which includes a link to these reports.

A few days before the accident, a specialized pipeline construction company began to conduct assembly works on a deflated and secured propylene pipeline route. The aim of the assembly works was to exchange several parts of the pipeline as a preventive maintenance measure. On October 17, 2016, a fire started at approximately 11:30 a.m. near the assembly works. An explosion occurred during the initiation of emergency operations, most likely at the ethylene pipeline. The explosion led to subsequent fires at various points along the pipeline trench, damaging further product and supply pipelines. Four people died, seven sustained serious injuries, and twenty-two others were injured in this incident.

The BASF facility is located in Ludwigshafen, Germany, officially called Ludwigshafen am Rhein. It is a city in the German state of Rhineland-Palatinate, situated on the river Rhine (Upper Rhine), opposite Mannheim. It is home to the world's largest integrated chemical complex. BASF is a sprawling chemical manufacturing site with approximately 2,000 buildings and a vast network of pipelines and rail tracks. The pipelines are used to transfer various chemical products to the production units and for shipping. Ludwigshafen is the headquarters of BASF, which has nearly 200 production facilities.

BASF reported that maintenance work was being carried out on the propylene pipeline on October 17, when the tragic incident occurred. A specialized pipeline construction company was replacing sections of the propylene pipeline when the fire started in the pipeline trench. Shortly after the emergency responders arrived and initiated fire response, an explosion occurred at approximately 11:20 a.m., likely on the ethylene pipeline adjacent to the pipeline where maintenance was underway. The explosion damaged adjacent pipelines, resulting in additional product release and the spread of the fires.

Local media reported that flames and a cloud of smoke could be seen in a harbor area north of the town, where oil and gas tankers supply the plant. Police quickly ruled out that the incident was related to a terror attack. A smoke cloud was seen high above the facility, and the harbor and nearby communities were alerted about potential air contamination. Several hours after the explosion, some smaller fires continued to burn, but all were under control. BASF reported that 14 operational facilities were shut down due to the incident, including two ethylene production units, known as steam crackers.

An investigation began almost immediately, and on Wednesday, October 26, the District Attorney of Frankenthal/Palatinate and the Police Headquarters of Rheinpfalz announced the following information on the progress of the investigation: "In the course of intensive investigations directly at the scene of the incident, it has been established that there was a cut made in a pipeline. This was apparently done with a cutting disc. Maintenance work using an angle grinder was in progress on an adjacent pipeline. The cut pipeline was not part of this work. This pipeline contained flammable raffinate" (Excerpt of the press release from the District Attorney of Frankenthal/Palatinate and the Police Headquarters of Rheinpfalz, translated by BASF).

Safety is a core value at BASF, and BASF also commissioned several additional investigations and studies, including the following:

- BASF has commissioned an independent expert to analyze the accident. The expert's report will analyze the sequence of events and the cause of the incident.
- BASF has commissioned a report by an independent expert for the safety-related assessment of the long-distance pipelines.
- Experts from the state criminal police are looking into the circumstances of the accident. Their investigations center on the maintenance procedures and the safety measures that were carried out prior to the maintenance work.
- Initial soil samples have been taken and provided to the criminal police. External experts will carry out further soil testing.

Author's Comments

- Detailed walkdown before work begins to ensure the work scope is understood
- Marking or otherwise identifying all of the equipment to be opened or cut

From initial reports, this seems to be a case of field workers accidentally cutting into the wrong pipeline, which had not been prepared for opening or maintenance, and appears similar to the incident that occurred at the Deer Park, Texas, PEMEX Refinery in October 2024 (based on preliminary reports from that incident). This underscores the need for clear communication between the permit issuer and the workers. Such communication should be established during a field walkdown between the permit issuer and the work crew before the permit is issued for the work to begin.

It also illustrates why it is important to clearly identify exactly which pipeline or equipment the work permit covers. One way that I have seen this done is by applying a colored paint stripe along the entire pipeline being prepared for maintenance to distinguish it from others that are nearby but may remain in service. This is an effective way to help prevent workers from getting onto the wrong pipeline. This does not replace a distinctive tag or other identification placed on each flange to be opened, indicating that the flange is safe to open.

Additional References

BASF Chemical Company, Update: Fire at the North Harbor in Ludwigshafen (December 2, 2016), https://www.basf.com/global/en/media/news-releases/2016/12/p-16-391.

BBC News, Ludwigshafen Fire: BASF Blast Hits German HQ, https://www.bbc.com/news/world-europe-37676974.

Incident 13: Loss of Containment Due to the Failure to Properly Identify the Equipment for Opening, or Workers Cutting or Opening the Wrong Equipment

Exxon Chemical Plant, Baton Rouge, Louisiana

A major propane release occurred when pipefitters opened a 4″ flange on a propane manifold. The open flange was on the pressurized side of a valve on a propane line that had been fully prepared for opening. The incident resulted in significant changes to line-breaking procedures.

There was no ignition or fire, no one was injured, and no equipment damage resulted.

Date unknown (before 1990)

This incident was briefly mentioned in Chapter 4, "When are We Most Vulnerable to These Hazards?"; what happened at our facility was similar to other incidents discussed in this chapter. The main difference was that we were very fortunate. In this case, the large, pressurized vapor cloud did not ignite. There were no fires, explosions, injuries, or equipment damage. However, we gained valuable lessons that were quickly incorporated into our line-breaking practices and procedures.

Unfortunately, I don't remember exactly when this incident occurred, but it was before 1990. Still, I remember it like it was yesterday. I was there as one of the responders, and I hope I never have to be involved in something like that again. The date isn't important; what matters is that I clearly recall the changes we made to our line-breaking procedures to prevent this from happening again. Thanks to those changes, I believe it shouldn't happen again.

I am sharing the key improvements to our practices and procedures to help prevent these incidents across the industry. The main changes we implemented are outlined below, but this is not a comprehensive list.

- The operations team is responsible for ensuring that all flanges to be opened are confirmed to be at zero energy before issuing the work permit. All vents and bleeders should be open and checked to be unobstructed, and all cross connections must be fully isolated.
- Each line or vessel opening task requires a joint site visit involving the permit issuer and those performing the work. If this isn't possible, a joint site visit must be held between the permit issuer and the craftsmen's representative. In this case, the representative is responsible for asking questions on behalf of the craftsmen and relaying all aspects of the job to them.
- The work permit is issued on-site at the work location. It is not issued in the control room or any other area away from the job site.
- The flange(s) to be opened are identified with a tagging system using a uniquely designed tag specifically created for this purpose. At our site, we required a weatherproof and tamper-proof, vividly colored tag that is easy to attach to the flange or equipment.

  Our operators quickly labeled this tag as the "crazy tag." The tag is intended to be easily recognized by craftsmen to identify flanges and/or access plates that are to be opened. To prevent flanges or access plates from being opened too early, these tags should only be used after the line has been fully prepared and confirmed to be at zero energy. The "crazy tag" is attached to each flange or access plate that is to be opened.
- During the joint job-site meeting, the proper tags are explained to the craftsmen, and they are instructed not to open any flanges or access plates unless they are correctly tagged.

Series III. Loss of Containment Due to the Failure to Follow Procedures or Noncompliance with Work Permit Requirements

Incident 14: Loss of Containment Due to the Failure to Follow Procedures or Noncompliance with Work Permit Requirements

Astron's Cape Town Refinery in Cape Town, South Africa, experienced a tragic reformer furnace explosion on July 2, 2020, when operators failed to follow procedures during a furnace restart. An investigation revealed that fuel gas and pilot gas valves were not properly closed before restarting the furnace. The incident resulted in two fatalities and seven injuries, all of whom were hospitalized.

This reformer furnace explosion occurred on July 2, 2020, while the reformer unit was being returned to service following a maintenance outage. A South African inquest found that operators, including a Shift Team Leader, were negligent by failing to follow procedures during the furnace restart, causing the explosion.

Details of this incident are a bit sketchy. However, Astron Refinery had just completed a planned maintenance turnaround and was returning units to service when the reformer furnace exploded. The explosion resulted in the deaths of two employees and injured seven others, who all required hospitalization.

The explosion occurred while refinery operators attempted to light burners in the reformer furnace. Magistrate Ingrid Arntsen, after reviewing evidence on this explosion and interviewing witnesses, ruled that the explosion occurred due to the negligence of five operators due to their failure to follow procedures, resulting in the explosion.

The Magistrate found that the standard safety procedures required that all pilot and fuel gas face valves be closed before lighting the furnace. According to her findings, the incident occurred after gas valves were not properly closed, allowing gas to enter a furnace during startup procedures following major maintenance. "All witnesses who testified as to the valves agreed that if the face valves were closed as required, the explosion would not have taken place."

The Magistrate reported, "The unit operators were negligent in failing to close the face valves as required ... and that negligence is causally linked to the explosion that killed the two deceased." The Magistrate also found the shift leader negligent for failure to wait for a backup colleague before lighting the furnace, in direct violation of the procedure. According to the magistrate, Astron Energy complied with its duty in terms of occupational health and safety laws.

The 100,000-barrel-per-day refinery was shut down for nearly three years due to the explosion and other plant repairs and modifications. The company has expressed its condolences to the families of those who were killed in this tragic incident.

I attempted to obtain permission to use photographs of the severely damaged furnace without success; however, they are available at the links in the "Additional References" section of this chapter.

Key Lessons Learned

The primary lesson learned in this incident is that detailed written procedures are so important, and even more important is that those procedures are closely followed in the field at all times, especially during transient operations such as a unit startups or shutdowns. The comments by the Magistrate are indications that these procedures were either less than adequate or, most importantly, were not followed in the field.

Author's Comments

Safety, Health, Environmental Procedures (SHE Critical)

I believe a critical procedure should be classified as a SHE critical procedure when omitting or skipping a step or executing steps out of sequence could cause personnel injury or equipment damage. Procedures related to returning a furnace to service, such as lighting furnace pilots or burners, would fall into this category. SHE Critical Procedures should require documented, step-by-step instructions with operator sign-off for each step as it is completed. This ensures that steps are not skipped and are performed in the correct order. Operators must be trained in all SHE Critical procedures, understand the significance of this designation, and strictly follow SHE Critical procedures in the field.

Lighting furnace burners also requires thorough purging of the firebox before attempting to light the furnace pilots or burners. Purging of the firebox is done to ensure that all potential flammable gases have been eliminated before igniting a flame inside the firebox. This process is usually a documented SHE Critical Procedure related to a furnace startup. It was not discussed in the Magistrates' report, and I am unsure if proper purging was performed before the explosion. Nevertheless, proper furnace purging should be completed for every furnace relight before introducing a flame into the firebox. This can be achieved by operating the fans prior to relighting for furnaces equipped with induced draft and forced draft systems. Otherwise, it is typically done by purging the firebox with high-pressure steam, followed by a gas test to ensure no flammable material is in the furnace.

Additional References

BIZ Community Health and Safety, Inquest Finds 'Negligence' Caused Astron Refinery Explosion, https://www.bizcommunity.com/Article/196/864/240365.html.

Creamer Media's Engineering News, Inquest Finds 'Negligence' Caused Astron's Cape Town Refinery Explosion, https://www.engineeringnews.co.za/article/inquest-finds-negligence-caused-astrons-cape-town-refinery-explosion-2023-07-21#:~:text=%22The%20unit%20operators%20were%20negligent,%2C%22%20an%20NPA%20spokesperson%20said.

News 24, WATCH | Two Killed, 7 Injured in Cape Town Refinery Explosion, https://www.news24.com/southafrica/news/watch-explosion-rocks-cape-town-refinery-structure-damaged-20200702.

Incident 15: Loss of Containment Due to the Failure to Follow Procedures or Noncompliance with Work Permit Requirements:

Chevron Pascagoula Refinery, Pascagoula, Mississippi

The Chevron Pascagoula Refinery experienced a brief electrical power outage, causing several pieces of equipment to trip offline. A process operator quickly attempted to restart a process unit furnace when it exploded, resulting in her death. OSHA cited Chevron for failing to have detailed furnace relight procedures that may have prevented this incident.

November 15, 2013; one fatality, zero injuries

The OSHA investigated this incident and reported the electrical failure and the furnace explosion. OSHA reported that the operator was lighting a series of burners in the furnace when it exploded.

When relighting a furnace after a trip, if no pilots or burners are still operating in the firebox, it is critical that the furnace undergo a thorough purge to remove any flammable gases before attempting to ignite a pilot or burner. Otherwise, remaining flammable gases in the firebox can cause an explosion when a torch or other flame is introduced. If pilots or burners are still lit and burning, there should be no need to purge the firebox before lighting other pilots or burners. Clear written procedures should be available for operators to safely return a furnace to service for unit startup, especially after a unit trip.

OSHA reported that Chevron did not make clearly written furnace procedures available to the operators to follow. OSHA reported that Chevron was using the same procedure that covers a unit startup, instead of specifically for startup after an emergency shutdown or unit trip. OSHA also reported that the operator did not purge the firebox before the attempted relight. See Figure 11.21 for a photograph of the Chevron Pascagoula incident.

Courtesy of Matt Defelice.

Figure 11.21 Photograph of the Chevron Pascagoula furnace explosion.

Author's Comments: Importance of Safety, Health, and Environmental Procedures

Furnace relight procedures are essential and should be organized in a checklist format. The operator must sign off on each step as it's completed to ensure no steps are missed and that they are performed in the proper order. This clearly includes the furnace purge procedures. After a furnace trip, especially if all furnace burners and pilots (if equipped) are extinguished, the purge removes any residual flammable gas from the firebox and helps prevent the risk of a furnace explosion when introducing a flame to relight pilots or burners.

Throughout my career, I've seen many incidents like this one. Clearly written procedures that follow a step-by-step process, requiring operators to sign off at each step and be strictly followed in the field, will save lives.

Additional Reference

US Occupational Safety and Health Administration (OSHA), Chevron USA Furnace Explosion, https://www.osha.gov/ords/imis/establishment.inspection_detail?id=954037.015.

Series IV. Loss of Containment Due to the Failure to Achieve Energy Isolation

Incident 16: Loss of Containment Due to the Failure to Achieve Energy Isolation

Tesoro Refinery, Martinez, California
Tesoro Martinez Sulfuric Acid Spill – February 12, 2014
February 12, 2014; zero fatalities, two sustained serious chemical burns

Two employees sustained serious sulfuric acid chemical burns while returning a sulfuric acid sampling system to service.

This incident was investigated by the CSB, and excerpts from the CSB final report are included in this summary.

On February 12, 2014, two employees at the Tesoro Martinez Refinery in Martinez, California, suffered first and second-degree chemical burns while working to put a sulfuric acid sampling system in the refinery's alkylation unit back in service following nonroutine maintenance.

The incident occurred when the operators opened a valve (Figure 11.22) to return an acid sampling system to service following maintenance. Shortly after fully opening the valve, the tubing immediately downstream of the valve detached at the connector, spraying the two operators with acid. The operators quickly proceeded to a nearby safety shower.

Until responders could isolate the system, acid continued to spray out of the failed tubing for over two hours from a nearly 100,000-gallon acid settler operating below 100 pounds per square inch gauge (psig) and near the ambient temperature of approximately 50 °F. This resulted in the release of approximately 84,000 pounds of sulfuric acid.

At the time of the incident, the two employees were wearing the personal protective equipment (PPE) required by company policy to put the sample system back in service – hard hat, safety shoes, goggles, flame-retardant jacket and coveralls, and gloves. This PPE was, however, insufficient to protect against an acid spray injury.

Emergency responders transferred the workers by helicopter and ambulance to the UC Davis Regional Burn Center for evaluation and treatment of their sulfuric acid burn injuries. Neither employee returned to work for over 150 days.

The CSB found that this second incident shared similar causation with a 1999 incident at the same refinery, then called the Avon Refinery, owned by the Tosco Corporation, that resulted in four fatalities. Similarities between the two incidents suggest that the Tesoro Martinez Refinery did not effectively continue to implement or communicate important safety lessons from the 1999 Tosco incident.

Figure 11.22 Photograph of the valve opened by operators, leading to the acid release. Note that photo annotations, and the following text is also courtesy of the CSB: "Two workers were injured in this acid release incident on February 12, 2014. This loss of containment from the open-ended tubing released 84,000 pounds of sulfuric acid."

The California Division of Occupational Safety and Health (Cal/OSHA) required the process to remain shut down from February 18, 2014, until February 28, 2014, when Tesoro completed abatement of the items identified in the order to prohibit use. The Cal/OSHA PSM Unit investigated the incident and issued citations totaling $51,450.28. The citations were later reduced to three general and one serious, for a total penalty of $43,400.

Author's Comments

- Workers' personal PPE should always be suitable for the potential hazards they may encounter
- Pressure testing after maintenance to confirm that the piping is fully connected and not leaking

This was a serious incident that resulted in significant and painful chemical burns. At many refining and petrochemical sites where acid is used, the toe wall surrounding the acid-containing equipment is painted a distinctive color, usually red. This serves as a warning to employees working in this area that acid is present in the equipment and that there is a risk of acid exposure. Workers are expected to wear protective gear when working in these areas, including impermeable clothing and gloves, goggles, and face shields.

It is unclear from this report whether the acid piping was pressure tested after maintenance to confirm it was fully connected and ready to return to service. This should be a standard procedure following any maintenance or new construction. Pressure testing must be conducted to verify that no connections are loose, disconnected, or leaking for any reason. In most cases, this can be done using water in a hydrotest, but sometimes steam or nitrogen at pressures near the operating level is used.

Additional References

US Chemical Safety and Hazard Investigation Board, Tesoro Martinez Sulfuric Acid Spill, https://www.csb.gov/tesoro-martinez-sulfuric-acid-spill/.

US Occupational Safety and Health Administration (OSHA), 29 CFR 1910.132, General Requirements (Includes PPE Requirements), https://www.osha.gov/laws-regs/regulations/standardnumber/1910/1910.132.

Incident 17: Loss of Containment Due to the Failure to Achieve Energy Isolation

Shell Singapore, Pulau Bukom Refinery, Singapore

Two different tasks were performed in proximity to each other. Hydrocarbons were drained to grade while preparing equipment for opening, resulting in ignition from adjacent hot work. Six workers were burned due to the fire, two seriously (three were hospitalized).

August 21, 2015

This incident data was extracted from the Singapore MOM and local media.

At this refinery on August 21, 2015, two separate tasks were underway on the crude distillation unit. One involved removing existing piping from the unit structure using an oxy-acetylene torch to burn the piping from supports while working from a scaffold. The second task involved workers on the ground removing piping joints from a hydrocarbon solvent piping circuit. This task required disconnecting a piping joint from a valve and attaching a drain hose to remove residual flammable hydrocarbons from the pipe into a nearby pit to contain the liquids. When one worker opened a valve on the solvent system to start draining, the liquid vaporized quickly, and the flammable vapors ignited from sparks caused by hot work overhead. Although the worker was promptly alerted and shut the valve, the fire rapidly erupted, directly impacting the workers on the scaffold.

As the workers escaped from the scaffold, six were exposed to the fire, resulting in injuries to all of them. Two were severely burned, with burns covering over 50% and 70% of their bodies. The Bukom Emergency Response Team contained and extinguished the fire within approximately 30 minutes. As a result of this fire, the crude unit was shut down for an extended period.

The Singapore Ministry of Manpower found that Shell Singapore had a "systemic failure in Shell's oversight to check for compatibility of work activities carried out simultaneously within the same vicinity. The hot works and cold works carried out by the two groups of workers in the same vicinity were not coordinated, thus creating a situation where flammable vapours (vapors) generated by the cold works were ignited by sparks from the hot works."

Er. Go Heng Huat, MOM's Director of the Major Hazards Department, said:

> "The refinery, as a major hazard installation, must properly manage safety and risk control measures. The lives of workers and the public could have been put at risk because adequate control measures were not properly implemented. Even though there was no loss of life in this case, the potential for more severe consequences was evident. MOM will continue to strengthen its enforcement of companies' workplace safety and health practices, including prosecution when there have been infringements that put the lives of workers at risk."

Shell management was also quoted in the local press as saying:

> "Since the incident in August 2015, we have extended our full cooperation to the relevant authorities, learnt from the incident, and implemented improvements across our site.
>
> "Shell has a goal zero ambition to achieve no harm and no leaks across all of our operations. We are committed to operating responsibly and safely, causing no harm to our staff and contractors. Incidents like this are unacceptable, and we are committed – at all levels of our organisation (organization) to continuously strive toward an incident-free workplace."

Author's Comments (Key Lessons Learned)

When preparing equipment for opening, we should aim to drain it into a closed system, such as a slop tank, to prevent the vaporization of lighter products, which can create a flammable atmosphere. Any draining of hydrocarbons should only be done in areas free of ignition sources. Typically, the work group performing the draining should conduct a walkaround before releasing the product to ensure there are no ignition sources nearby, such as burning, welding, or other arcing or sparking devices in use.

This is especially true when two or more work teams work nearby. In these cases, clear communication and planning between the groups are essential to ensure coordinated work activities and prevent impacts like those in this incident.

Additional References

Oil and Gas Journal, Shell Issues Force Majeure After Upset at Singapore Ethylene Complex, https://www.ogj.com/refining-processing/petrochemicals/article/17246254/shell-issues-force-majeure-after-upset-at-singapore-ethylene-complex.

Oil and Gas Journal, Unit Remains Shuttered at Shell's Pulau Bukom Complex, https://www.ogj.com/refining-processing/refining/operations/article/17245072/unit-remains-shuttered-at-shells-pulau-bukom-complex.

Singapore Ministry of Manpower, Company Fined $400,000 for Fire at Petroleum Refinery in Pulau Bukom, https://www.mom.gov.sg/newsroom/press-releases/2019/0108-company-fined-for-fire-at-petroleum-refinery-in-pulau-bukom.

The Straits Times, Shell Fined $400,000 over Pulau Bukom Fire that Injured Six Workers, https://www.straitstimes.com/singapore/courts-crime/shell-fined-400k-over-pulau-bukom-fire-that-injured-six-workers.

Incident 18: Loss of Containment Due to the Failure to Achieve Energy Isolation

Valero Oil Refinery, Memphis, Tennessee

Three workers were in the process of installing isolation blinds in the 36-inch outlet from the refinery flare to the flare knockout drum when a flash fire erupted. One worker was killed, and two others were seriously burned. This incident was investigated by Tennessee OSHA (TOSHA). This summary is taken from the TOSHA report.

March 6, 2012; one fatality, two serious injuries

The Valero Memphis Refinery is located on 250 acres along the Mississippi River on Lake McKellar. It is a 195,000 bbl/day (BPD) refinery and produces primarily clean products such as gasoline, diesel, and jet fuels. The refinery has a good safety record and is recognized as an OSHA Voluntary Protection Program (VPP) Star Site and an American Fuel and Petrochemical Manufacturers (AFPM) Elite Gold Award recipient.

On March 6, 2012, three contract workers were working between a knock-out drum and the South Flare tower, opening a flange to install a 36-inch blind to isolate the knock-out drum from the flare system. As the contractors were working and the flange was opened, an explosion occurred, followed by a fire, which resulted in the death of one of the contractors and severely injured the other two individuals.

As a result of this incident, the Tennessee Occupational Health and Safety Administration (TOSHA) issued violations to Valero for 16 serious alleged violations and listed the potential cause as "failure to isolate and control ignition sources and allowing air into the piping that allows ignition." The investigation revealed that the workers were not protected from the dangerous gases in the flare knockout drum and flare piping. Also, the investigation confirmed that the pilot remained in service during the mechanical work. Among the citations, TOSHA said that Valero failed to protect the workers by failing to secure the piping by LOTO and to control the hazards by implementing energy control procedures. TOSHA also cited Valero for failure to provide protective clothing for the workers to help protect them from the hazards.

Author's Comments: The Importance of Equipment Isolation and Equipment Preparation

In this book, I emphasize the importance of equipment isolation and preparation before opening or entry. This involves double block and bleeders, when feasible, or at the very least, isolating the equipment from other live equipment, draining all liquid products from the equipment, and purging it to eliminate any remaining hydrocarbons or chemicals prior to turning it over to maintenance for equipment opening. I also highlight the necessity of removing all ignition sources from the area; in this case, the flare pilot was still in service when the equipment was opened, likely serving as the ignition source.

Also, all flanges or equipment to be opened must be adequately identified with a tag or other marking or identification that the flange is safe to open. This should be covered in detail during the equipment walkdown before the work permit is issued, so there is no confusion about what is safe to open.

This tragic event occurred at a refinery with an excellent safety record at the time of this event. This says that none of us can be complacent, even for an activity that, at first blush, appears to be remotely located from the major process units with minimal obvious safety hazards. We must use all our safety management processes and safety procedures on every task, every time.

Additional Reference

State of Tennessee/Department of Labor and Workforce Development, Citation and Notice of Penalty, Valero Memphis, Tennessee Refinery, https://nxstrib-com.go-vip.net/wp-content/uploads/sites/18/2012/09/valero.pdf.

Incident 19: Loss of Containment Due to the Failure to Achieve Energy Isolation

Chevron Pembroke Refinery, Pembrokeshire, Scotland

An explosion and fire occurred in an amine regeneration unit in one of the amine storage tanks. The explosion occurred during a routine tank cleaning operation in preparation for maintenance.

June 2, 2011; four fatalities, one serious injury

Incident Background

On June 2, 2011, four people were tragically killed and a fifth seriously injured when a tank exploded at Chevron's Pembrokeshire Refinery. This incident occurred at the refinery's Amine Regeneration Unit, where one of the amine storage tanks had been taken out of service and was being prepared for inspection and maintenance. At the time of the explosion, Tank 17T302 was in the process of being emptied with residual product, primarily diethanolamine (Amine), light naphtha, and diesel, prior to cleaning and gas freeing for entry. The explosive force resulted in the 5.5-ton (5.5-metric tonnes) tank roof being blown clear and landing 180 feet (55 m) away. The roof narrowly missed a multi-fuel pipe track and collided with a butane storage sphere, causing structural damage to the vessel legs and supports.

This tragic incident was investigated by the HSE and a detailed report was written, including recommendations for preventing a recurrence of an incident like this. This is a summary of the HSE report. See the link in the "Additional References" section of this chapter for a full copy of this report. Following this incident, the HSE took enforcement action to ensure that those responsible for the incident were held to account in a criminal court. Two defendants pleaded guilty to charges under Sections 2 and 3 of the Health and Safety at Work Act 1974. When passing sentence on the defendants at Swansea Crown Court on June 6, 2019, the Judge, the Honorable Mr. Justice Lewis, stated: "The fact is that the serious nature of the failings, the number and range of failings, and the length of time over which certain of the failings occurred demonstrate serious errors and serious failures within the organization to address the risks to health and safety."

Purpose of Amine Treating

Crude oil and intermediate refinery products consist of hydrocarbon compounds and other contaminants, one of which is hydrogen sulfide (H_2S). H_2S is corrosive to metals and is also toxic, and must be removed before the final products are sent to the consumer. This process relies on the use of a chemical, diethanolamine (amine), which has an affinity for hydrogen sulfide. The lean amine, amine that has not contacted hydrogen sulfide, is circulated to the intermediate product scrubber vessels, where the amine contacts the H_2S-bearing products. The hydrogen

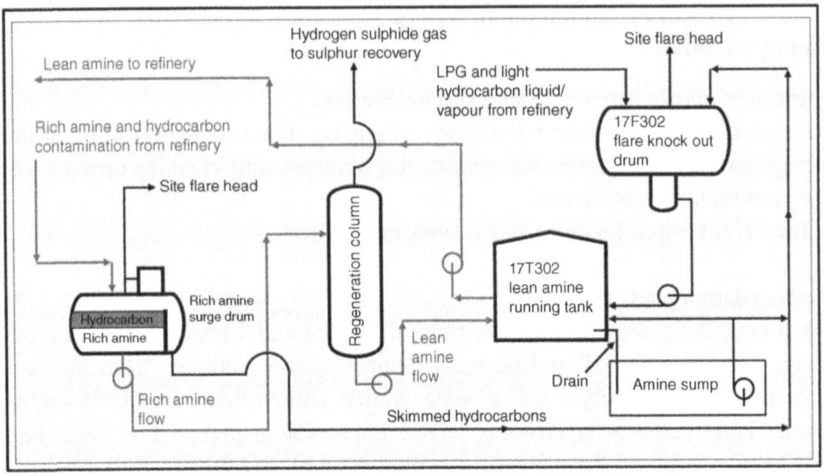

Figure 11.23 Simplified flow schematic for a typical petroleum refinery amine regenerator. Courtesy of the British Health and Safety Executive (HSE).

sulfide molecules quickly bond to the amine and are carried away with the foul amine, amine containing the hydrogen sulfide, to the amine regenerator. In the amine regenerator, the hydrogen sulfide is stripped away by heat and steam and is carried overhead to the sulfur plant, where it is removed as elemental sulfur, a saleable product. The recovered amine, now lean amine again, flows back to the scrubbers to continue the process. See Figure 11.23 for a simplified flow schematic for a typical amine regenerator.

Amine Scrubbing and Amine Regeneration

This amine process is designed to remove the hydrogen sulfide and other sulfur compounds from the product streams before the products go to the consumer. The amine scrubbers are physically located on the individual process units where the products contact the amine and where the hydrogen sulfide is absorbed by the amine. The amine regenerator is located separately and is where the hydrogen sulfide is stripped from the foul amine, allowing the lean amine to be recovered and returned to the process to continue the operation. The amine regenerator section contains the rich amine surge drum, the regeneration column, two vertical process tanks, identified as 17T302 and 17T303, and the flare system.

The two tanks were cone-roof design. 31.8 feet (9.7 m) in diameter and height, and were equipped with pressure/vacuum vents. Refer to Figure 11.24 for photographs of the two storage tanks. Although the tanks were used interchangeably, the original design was for two different purposes. Tank 17T303 was used for the

Courtesy of British Health and Safety Executive (HSE).

Figure 11.24 Photograph of the amine regenerator storage tanks prior to the explosion.

separation of waste streams, while 17T302 was designed to operate as the lean amine tank. Tank 17T302 was designated as the "running tank," kept at 104 °F (40 °C) and used for the storage and recirculation of lean amine to the refinery scrubbers. Tank 17T303 was used for the separation of "slop" material into its separate components. Both tanks contained a layer of about 1 foot (300 mm) of diesel to prevent oxygen ingress and to minimize the oxidation of the amine.

Hydrocarbon Accumulation

During scrubbing operations, the amine primarily binds to the hydrogen sulfide; however, a relatively small amount of hydrocarbons, including light hydrocarbons, are also carried back to the regenerator with the foul amine. The bulk of the hydrocarbons returning to the ARU in the rich amine were relatively volatile, within the range of propane to nonane. Over time, this hydrocarbon would accumulate in the lean amine tank in a layer above the amine. The layer would periodically reach a depth of about several feet (as deep as 3 feet) (over 900 mm). Periodically, this layer would be skimmed off using an internal weir and transferred to the flare system, where most of it would vaporize to the flare. The ARU flare drum received light hydrocarbon material from various sources, not just the ARU. The heavier molecules would accumulate in the flare boot, the bottom of the flare knockout drum, where they could be transferred to a refinery slop system.

However, in 1998, the decision was made to operate the ARU as a "closed system," which meant that the hydrocarbons and other residue would be pumped back to the running tank. As an important consequence of this change, any liquid hydrocarbon transferred to the ARU flare drum as a result of normal ARU operation, or events elsewhere on the refinery, was retained on the unit and eventually accumulated within the ARU running Tank 17T302.

Preparation, Draining, Isolation, and Testing of 17T302 Prior to Turnover to Maintenance

Draining of 17T302

A plant operator was assigned to drain Tank 17T302, a process with which he was not familiar, on May 18, 2001. He began this process by opening the main drain valve 1 (Refer to Figure 11.25 for a photograph of storage tank 17T302, with bottom drain details). The operator was not aware that the tank contained a bottom sump and siphon system and didn't realize that the tank could have been

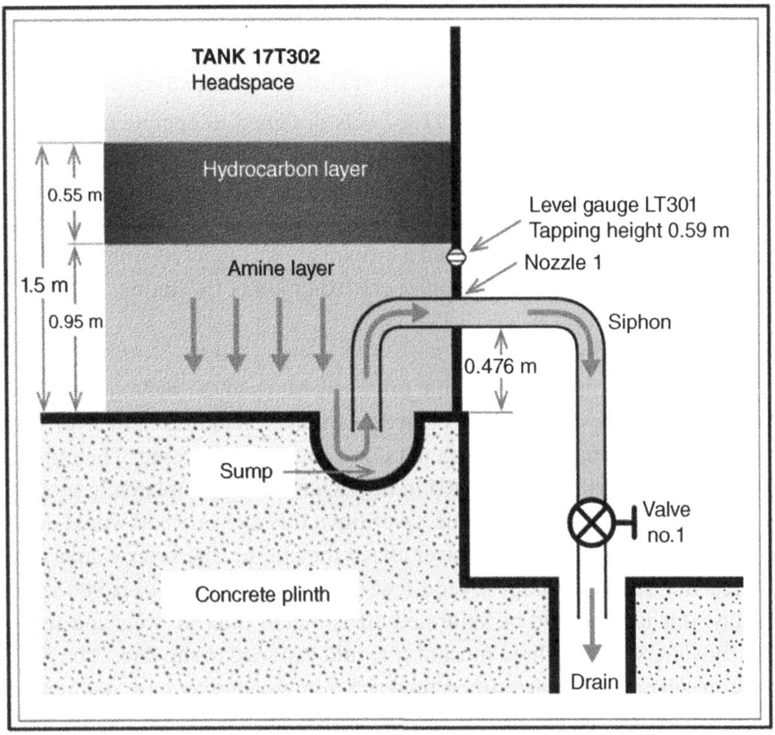

Figure 11.25 Photograph of storage tank 17T302, with bottom drain details. Courtesy of the British Health and Safety Executive (HSE).

completely emptied had this system been used properly. He drained the tank until he lost "siphon" and assumed the tank was empty at that point.

The operator was even more convinced that the tank was empty since he confirmed this reading on the digital display panel in the control room. The level as observed on the plant operator panel was reading 0.7 m (2.3 feet). The operator was not aware that the tank gauge could read the tank level down to the floor, and that a significant level of liquid remained in the tank.

Unfortunately, the operator recorded the tank level as being empty, even though it still contained a significant liquid level. This logbook entry was carried forward from shift to shift and was not questioned by others.

Isolation of Tank 17T302

Prior to tank entry, cleaning, and maintenance on Tank 17T302, contract personnel were provided with a work permit to isolate and blind all connections to the tank to isolate it from other process equipment in the ARU.

The permit was issued, and work was carried out during the period May 26–28, 2011. Isolation blinds (a flat metal plate) were installed in the pipe flanges to ensure positive isolation of each piping flange connection immediately at the tank. The craftsmen found significant liquid flow at each of these connections when the flange was opened, clearly indicating that the tank had not been emptied. For example, when opening the flanges, especially for the larger pipe connections, the HSE reported that the workers were met with a significant process flow at each flange. The workers reported a flow of hydrocarbon liquids with a visible haze on the liquid surface and a strong petrol-like odor, indicating the presence of light hydrocarbons. Refer to Figure 11.26 for a photograph of a typical isolation blind installed in Tank 17T302. Figure 11.27 illustrates the tank piping and shell manway elevations as compared to the liquid level in the tank while it was being blinded.

As it turns out, the work permit background information was assembled by a Chevron employee who had never previously prepared tank permit documents. He was able to identify amine and hydrogen sulfide as anticipated hazards within the vessel, but was unaware of the presence of diesel, light hydrocarbon, or potentially pyrophoric material.

Despite that tank not being fully empty, the work to isolate the tank from other ARU equipment continued. Apparently, no one suggested that work should cease, no one notified the process organization that significant liquid was spilling from the flanges, and that a significant amount of liquid remained in the tank. No one reported the presence of liquid light hydrocarbons and the significant fire and explosion hazard that existed at the tank with this work continuing. Instead, the isolation work continued.

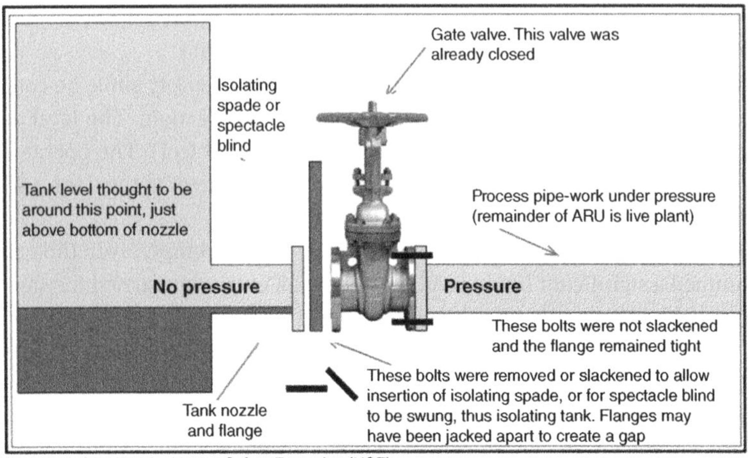

Courtesy of British Health and Safety Executive (HSE).

Figure 11.26 An example of isolation of piping flanges on storage Tank 17T302, showing the actual level in the storage tank at the time blinds were being installed. Note that the actual tank level was above the bottom edge of the piping flange for each of the piping connections. This resulted in the release of a significant amount of amine and hydrocarbon liquids when each flange was parted for blind insertion.

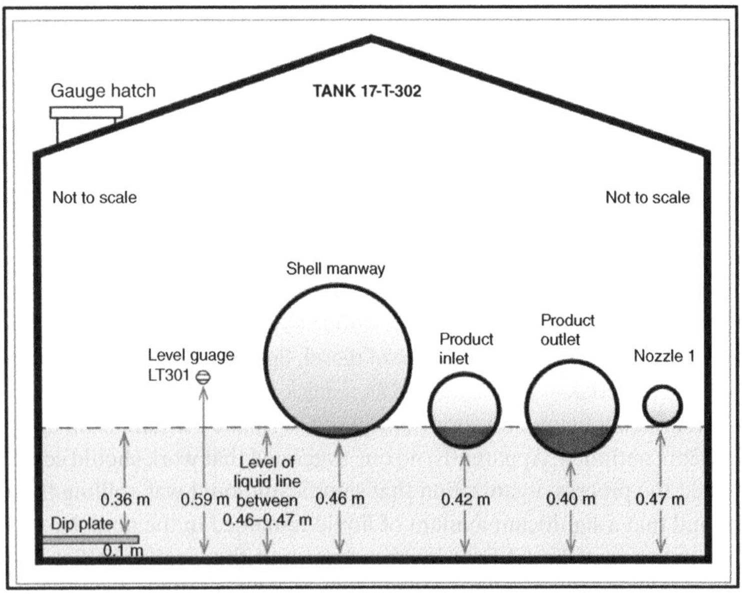

Figure 11.27 Example of the piping and shell manway elevations as compared to the liquid level in storage Tank 17T302. Courtesy of the British Health and Safety Executive (HSE).

The following day, the contract team returned to the site to continue the isolation work. They removed the access manway only to find the liquid level about 1 inch (2.5 cm) above the bottom lip of the manway bottom flange. A significant amount of liquid spilled from the tank and continued for about 45 minutes, until the manway cover was reinstalled. At this point, the workers became convinced that the tank had been emptied after all, but that amine had backed into the tank through a leaking isolation valve. This was recorded in the logbook and, unfortunately, was then carried forward from shift to shift. The HSE concluded this was not correct and that the tank had never been totally emptied of product.

The HSE reported that the operators appeared unaware that the sump and siphon drainage equipment could be used to empty the remaining liquid tank contents. What evolved was an informal plan to remove the remaining liquid using a vacuum tanker with a hose through the roof manway. There was no consideration of tank ventilation, bonding and grounding, or other safety considerations for remove the liquid. For example, no consideration was given to using portable pumps, or siphon from Valve 1, or for the introduction of water into the tank to reestablish siphon. The HSE reported that there was a facility to connect a vacuum tanker or diaphragm pump directly to the nozzle 1 drain manifold in order to remove the residual material from the tank; however, it appears that they went to the overhead removal without further consideration. Figure 11.28 illustrates the isolation spades and blinds used to isolate Tank 17T302.

Manual Dip Gauge and Gas Test

To confirm the hazardous nature of the tank atmosphere above the liquid level, the acting operations Advisor requested on May 31, 2011, that the plant operator carry out a manual dip gauge and a gas test on Tank 17T302. The operator conducted the testing through the tank roof gauging hatch using handheld portable equipment made for this purpose. The operator confirmed a liquid depth of 14.2 inches (36 cm) and noted this in his shift report. Unfortunately, the operator did not compare the manual dip reading with the computer panel level gauge. The HSE discovered that the dip reading was inaccurate and that it actually understated the level in the tank. The issue is that there is a 3.8 inch (9.6 cm) metal "guard plate" welded to the floor directly under the gauging hatch, designed to protect the tank bottom from repeated gauging. The plant operator was not aware of this plate and did not adjust the recorded level to obtain a true reading of the liquid depth, which was approximately 18.1 inches (46 cm). This places the actual tank level well above the bottom flange on all piping flanges, and slightly above the bottom of the access manway lower flange.

Courtesy of British Health and Safety Executive (HSE).

Figure 11.28 Photograph of isolation spades and blinds from Tank 17T302.

The plant operator, while never having conducted a gas test on a process vessel, conducted a gas test on Tank 17T302. He used a four-gas test machine capable of testing oxygen (% O_2), carbon dioxide (ppm), hydrogen sulfide (ppm), and hydrocarbons (% LEL), and took only one sample using a 1–1.5 m (5 feet) sample hose. The HSE makes the point that a well-trained gas tester will take several readings to ensure a representative sample of all sections of the tank. Samples should be taken at different heights, as the concentration levels within space may vary significantly.

The plant operator observed a hydrocarbon reading of 67% LEL and a hydrogen sulfide concentration of 10 ppm. He did not formally record the results within his shift report or elsewhere. However, he radioed the unit control room and discussed the "high" %LEL with his lead operator. During this discussion, the plant operator did suggest removing the drain isolation to empty the tank further, but no meaningful consideration was given to his suggestion.

From my perspective, a reading of 67% LEL is noteworthy and should have been reported broadly. All work at the tank should have ceased immediately. This is clearly well above the normal maximum of 10% LEL and should have been a red flag to all personnel on this job as a clear and immediate safety concern. The HSE also reported this.

Table showing gas test data taken by the HSE directly from the gas-test machine used to sample Tank 17T302.

Hydrogen sulfide (ppm)	Carbon monoxide (ppm)	Hydrocarbon (% LEL)	Oxygen (%)
10	213	67%	20.9

Courtesy of British Health and Safety Executive (HSE)

Author's Comments

The initial gas test indicating 67% LEL should have been a "red flag" warning to all personnel. All work should have been stopped, the area should have been evacuated, the vessel should have been re-purged, a new gas test conducted, and a new work permit issued before any work is resumed. The vessel should be below 10% LEL before work is resumed.

Liquid that was found "trapped" in the piping and vessel and draining to the ground should have served as a "red flag" warning to all the workers that the work should have been immediately stopped, the flange(s) should have been retightened, and the work area should have been immediately evacuated. A full risk assessment should have been completed, and the piping and vessels should have been completely drained or otherwise emptied, a new gas test conducted, and the work permit reissued before resuming the work.

Additional References

British Health and Safety Executive (HSE), Chevron Pembroke Amine Regeneration Unit Explosion, June 2, 2011, https://www.hse.gov.uk/comah/assets/docs/chevron-pembroke-report-2020.pdf.

British Health and Safety Executive (HSE), Explosion and Fire: Chevron Pembroke Refinery, June 2, 2011, https://www.hse.gov.uk/comah/chevron-refinery.htm.

IchemE, Lessons Learned Database, Lean Amine Tank Explosion – Pembroke (Pembrokeshire, Wales, June 2, 2011), https://www.icheme.org/media/26306/pembroke-incident-summary-02-jun-11.pdf.

Occupli, Rebecca Davis, May 30, 2024, Blog, What Happened in the Chevron Pembroke Refinery on June 2, 2011? https://occupli.com/what-happened-in-the-chevron-pembroke-refinery-explosion/.

Western Telegraph, June 2, 2011 – The Day the Chevron Refinery Blast Killed 4, https://www.westerntelegraph.co.uk/news/23562606.june-2-2011---day-chevron-refinery-blast-killed-4/.

Incident 20: Loss of Containment Due to the Failure to Achieve Energy Isolation

ExxonMobil Singapore Refinery, Jurong Island, Singapore

A fire occurred on an operating crude unit while blinds were removed from piping associated with an atmospheric fractionator overhead exchanger.

April 4, 2007; three fatalities and one injury.

This incident was investigated by the Singapore Civil Defense Force (SCDF). I am familiar with this incident, but I do not have permission to use photographs or other information directly from it. Therefore, this report is based mostly on local news reports, other easily accessible online sources, and my knowledge of crude unit operations. The incident occurred while the crude unit was operating normally; however, a section of the atmospheric crude unit overhead preheat circuit was out of service for maintenance. This report explains what happened and what could happen, and to help prevent a similar incident at another petroleum refinery or petrochemical plant worldwide.

This tragic accident happened around 1:15 a.m. on April 4, 2007, during the removal of isolation blinds from the related piping circuits after replacing an overhead crude preheat exchanger tube bundle. The planned maintenance was almost finished, and the isolation blinds, flat metal plates installed between the flanges, were being removed to prepare for returning the preheat exchangers to full service.

The exchangers were part of a set of "stacked" exchangers mounted in the structure next to the atmospheric crude fractionation column. These are typically very congested structures, usually with two or three overhead condensers operating in parallel directly above where the contractors worked, and often a bank of nine (sometimes more) atmospheric upper-sidestream exchangers arranged in a three-wide by three-high layout. These exchangers transfer heat from an atmospheric fractionator sidestream, generally light naphtha from an upper sidestream, to the incoming crude oil.

Several additional stacked exchangers exchange heat directly below these exchangers, usually from a mid to lower fractionator pumparound circuit with the incoming crude oil. The cooled pump-around stream is then returned to the lower fractionator to control the fractionation temperatures. The mid or bottom pumparound exchangers operate at very high temperatures, possibly above the autoignition point of the light naphtha from the atmospheric sidestream circuit. See Figure 11.29 for a sketch of a typical stacked exchanger train associated with a crude unit atmospheric column.

This illustration does not depict the actual unit discussed here. It shows a typical atmospheric crude unit stacked exchanger train. It aims to illustrate how crowded this area is for workers to isolate and install or remove blinds in the connecting

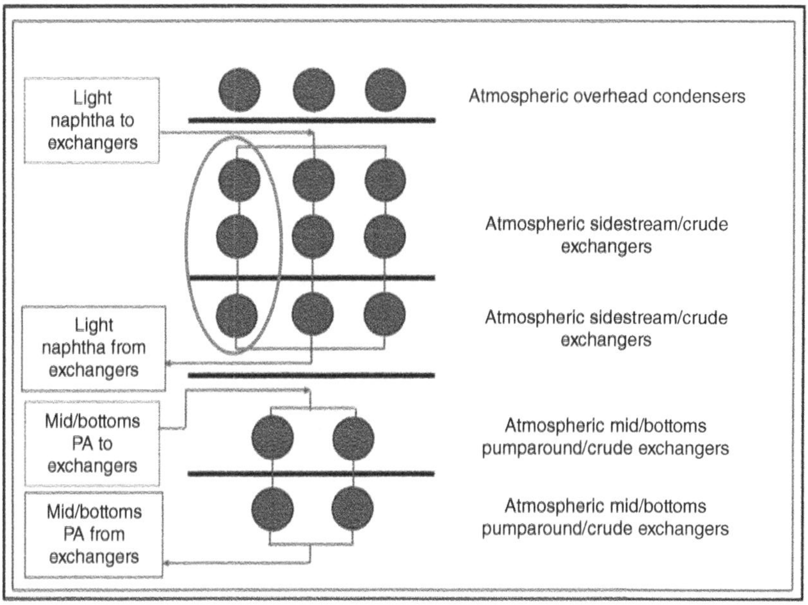

Figure 11.29 Sketch of a typical atmospheric crude unit stacked exchanger train.

large-diameter piping. If you imagine scaffolding erected around the structure to provide access to the pipe flanges, you'll understand how congested this area is. Additionally, all exchangers are in operation except those being isolated; therefore, it would be very hot and uncomfortable to work around and between the operating exchangers. The three exchangers enclosed by the oval indicate the units where deblinding occurred.

The planned work involved isolating one bank of the atmospheric sidestream exchangers to facilitate the removal of the heat exchanger bundles for maintenance and cleaning. Isolation included closing the valves on the exchanger inlet and outlet piping, draining or removing remaining hydrocarbons, and installing blinds (a flat metal plate) in the exchanger inlet and outlet flanges to ensure complete isolation. The remaining two banks of the atmospheric sidestream heat exchangers remained in service while this work was completed.

All of this was completed successfully, and the work at the time of the accident involved deblinding or removing the isolation blinds to restore the system to service. Reportedly, five contract personnel were assigned to this exchanger bundle cleaning and replacement job, working on a scaffold around the three exchangers and removing blinds from the related piping systems. As the discussion above shows, this is a very congested work area with multiple heat exchanger decks and large-diameter piping on both the tube side and shell side, connecting the exchangers to each other. To make things more complicated, scaffolds must be erected to access the isolation blinds, which further restricts access and egress from the scaffolds and platforms.

What is known about this incident is that the release of light naphtha occurred when the flange for the last blind was being opened on the atmospheric sidestream connection to the exchanger. Reportedly, when the flange was opened, naphtha drained from the flange. Reportedly, workers responded by using a water hose to wash the naphtha from the deck as the flange continued to release product. The released naphtha and water cascaded down through the structure onto the hot bottoms pumparound exchangers located near grade. While not certain, the most likely source of ignition would have been the bottom pump-around exchangers, due to their very high operating temperature. Light naphtha on a crude unit upper sidestream is highly volatile. It is typically a gasoline boiling range material with an extremely low flash point and a mid-range autoignition point.

When the ignition of the naphtha occurred, the fire raced up through the structure where the contractors were working. As a result of this tragic fire, two of the contractors were killed instantly, and a third died later at the hospital. The fourth contractor sustained severe burns to one of his legs but survived.

Author's Comments

I know you are familiar with the old saying that "hindsight is 20-20," but if I were investigating this incident, I would be seeking the answers to the following questions. Of course, to help prevent a recurrence of this kind of incident, this should be applied to all work involving OPE:

- Was there an operator assigned and present in the structure while the deblinding was being done?
- Were all the isolation points double-checked to verify that there were no hydrocarbons present before the work permit was issued for the deblinding?
- Were the vents and drains on the piping systems to be opened in the fully open position and car-sealed open and unobstructed?
- Were the scaffolds built with emergency egress to allow the workers an opportunity to escape in the event of an emergency?
- Were the contractors fully qualified for the work they were doing?
- During the worksite job visit, were the contractors reminded to quickly reclose a flange if pressure or liquid was found?
- Was there adequate fire response equipment prepositioned at the site where flanges were being opened?
- Were the workers wearing H_2S detection devices while opening flanges on the light naphtha circuit?

Why Are These Questions Important?

Sometimes, when it comes to deblinding, it's easy to think of this as the least risky part of the process. After all, the planned work is mostly done, and we are only removing the isolation blinds to put the equipment back into service. However, this incident clearly shows that this is not the case.

The truth is that valves can leak, and liquids can build up in areas where we might not expect them. That's why the joint job-site visit is so important. During this visit, we can check all the vents and drains, ensure they aren't blocked, and confirm that the vents and drains are fully open with no product leakage.

As mentioned earlier, all vents and drains on the equipment that need to be opened should be verified as fully open and securely sealed in the open position. Valves are known to leak, which can allow hydrocarbons or chemicals to accumulate in the piping systems or directly against the blind being removed. If the vents and drains are completely open, or if they are opened during the joint job-site visit, any leakage will be quickly detected, preventing a major loss of containment. It is also good practice during the joint job-site visit to have clear communication between the work permit issuer and the permit receiver. This should include an on-site discussion clearly identifying exactly where each flange is to be opened and how the flange is identified, as well as confirming that zero energy has been achieved – showing the permit receiver where the vents and drains are fully open.

It is good practice for work permit procedures and practices to require an operator to be present when the flanges or access plates are opened. This applies both during the initial opening to install the isolation blinds and when the work permit is issued to remove the blinds. Both situations involve opening flanges to install or remove the blinds. The operator's presence ensures that workers are opening the correct flange(s), following proper procedures, such as cracking each flange before fully removing all the bolts, and always breaking the flange away from them to prevent anyone from being in the line of fire. The operator understands the process and the hazards involved. They are familiar with the characteristics and properties of the process material and the precautions needed when working with it. Additionally, if pressure or liquid is present, they can ensure the flange is immediately closed to contain the release and prevent spills.

Whenever possible, scaffolds should be constructed to fully comply with OSHA scaffold requirements, known as a green tag scaffold. A green tag scaffold is built with proper guard rails, toe boards, and meets all access and egress standards. Typically, a scaffold built to these standards does not require fall protection, which could hinder emergency egress, but it still provides emergency egress, even in heavily congested areas.

Employees and contractor personnel should be trained in the applicable safety management procedures and practices. These requirements are included in OSHA regulations as outlined in the following paragraphs, which detail the company's and the contractor employer's training development and implementation responsibilities. Although Singapore is not under OSHA jurisdiction, I have found that these regulations serve as a valuable guide to follow, regardless of jurisdiction. Note that this is a summary; refer to the regulation for more details.

> The employer (the company) is responsible for developing and implementing process safety management work practices for employees and contract personnel (29CFR 1910.119(f)(4)).
>
> The contract employer shall ensure that each contract employee is trained in the work practices necessary to safely perform his/her job, that the employee is instructed in potential fire, explosion, or toxic release hazards, has received and understood the training, and prepares a record of the training. The contract employer shall ensure that each contract employee follows the safety rules of the facility, including the safe work practices required by the employer (the company) (29CFR 1910.119(h)(3)(i) – 1910.119(h)(3)(iv).

OPE is a higher-risk task with the potential to encounter flammable, toxic, or pyrophoric materials. A good practice is to have at least one fully charged high-pressure water fire nozzle readily available in case of a loss of containment or a flash fire. The nozzle should be positioned outside the hot zone but kept accessible as a precaution.

Other Considerations When Opening Process Equipment

Personnel opening equipment containing or that has contained hydrogen sulfide may face a higher concentration of H_2S when opening flanges or other process equipment. It is advisable, and possibly required, for these personnel to wear an H_2S detector, and they should be trained to leave the area immediately if the alarm sounds. Alternatively, workers can wear hose-line supplied breathing air or an SCBA.

All flanges or access plates that need to be opened should be tagged with a uniquely identified label, designed and used exclusively for OPE tasks. This label should be attached by the operating personnel directly to the flanges or access plates only after the equipment has been depressurized, drained, and purged to remove any residual hydrocarbons or chemicals.

Ensure proper signage and barricades are in place, alerting others that OPE work is in progress and that personnel access is limited to only those directly involved in the work.

A drain pan should be provided to safely collect any potential spillage from each flange or access plate that is opened. Always be aware of the risk of static electricity, and ensure the drain pan is conductive, properly grounded, and bonded to the equipment being drained.

Personnel performing the work should be reminded that if a flange leaks when opening, they should immediately reclose the flange to prevent the product from spilling.

Additional References

ExxonMobil Corporation, Gasoline SDS, https://public.spheracloud.net/Pdf/MsdsIAImages/Zy92XWn6Bkq2c41K6eIc0g.PDF.

US Occupational Safety and Health Administration (OSHA), Standard Interpretations Employer Responsibilities and Contractor Responsibilities Under the PSM Standard, https://www.osha.gov/laws-regs/standardinterpretations/2022-02-16#:~:text=The%20contract%20employer%20remains%20responsible,1910.119(f)(4).

US Occupational Safety and Health Administration (OSHA), By Standard Number 1910.119 – Process Safety Management of Highly Hazardous Chemicals, https://www.osha.gov/laws-regs/regulations/standardnumber/1910/1910.119.

Wild Singapore, Channel NewsAsia, May 3, 2007, http://www.wildsingapore.com/news/20070506/070503-7.htm#:~:text=SINGAPORE:%20An%20early%20morning%20fire,the%20cause%20of%20the%20fire.

Wild Singapore, News Homepage, https://wildsingaporenews.blogspot.com/2011/01/exxonmobil-refinery-fire-seven-charged.html.

Wordpress/The Straits Times, Migrant Workers Singapore, an Archive of News Stories Related to Migrant Workers in Singapore, https://migrantworkerssingapore.wordpress.com/2007/05/.

Incident 21: Loss of Containment Due to the Failure to Achieve Energy Isolation

Giant Industries, Cinza Oil Refinery, Jamestown, New Mexico

A major loss of containment and fire occurred during the opening of the alkylate product pump for maintenance.

April 8, 2004; six people were injured, including four with serious burn injuries

This alkylation unit experienced a major loss of containment of light naphtha and a major fire, which caused six injuries, including four serious burn injuries. The incident occurred due to the incomplete isolation of a product pump that was being opened for maintenance. This incident was investigated by the CSB, and a thorough report is available in the "Additional References" section of this chapter.

What is provided here is a summary of this incident. See the full report for additional details.

While opening flanges on an alkylate product pump, a highly flammable gasoline component ignited, causing a loss of containment and a major fire. This incident happened on April 8, 2004, at the Giant Industries Ciniza Refinery near Gallup, New Mexico. Four workers were seriously burned due to exposure to the hot liquid and the resulting fire. The incident happened at the refinery's HF Alkylation unit, although no HF was involved in this incident.

The workers were opening flanges on the pump, which was thought to be isolated, drained, and vented at the time. Unknown to the workers, one of the pump's suction valves, used to isolate the pump, was in the fully open position. The pump was fully pressurized with hot and volatile alkylate, resulting in the release of the hot product and a significant fire.

Two other workers received minor injuries, and the incident caused the evacuation of nonessential employees as well as customers of a nearby travel center and truck stop. The fire caused damage to the surrounding refinery equipment and support structures. As a result of this incident, it was reported that the unit was down until late in the fourth quarter, and the damage was estimated to be in excess of $13 million.

The day before this incident, operations removed the pump from service and switched to the steam turbine-driven spare pump. They closed the isolation valves, including the suction valve to the pump, and tagged each valve, indicating the pump was out of service for maintenance. They connected a drain hose to a bleeder on the pump and vented it to depressurize and drain the remaining liquid. The operators, as they frequently did, used the position of the valve handle to confirm that the valves were fully closed. The type of valve involved was a rotating plug quarter-turn valve operated by a single lever or bar. It was also equipped with valve position indicators, which are more difficult to read but are designed to show the actual valve position. Refer to Figure 11.30 for an image of the suction valve with the operating lever in the perpendicular position.

As it turns out, this suction valve had been overhauled in an outside shop in the past, and when returned, the position of the lever or bar had been reinstalled 90° from its original position. Using the valve handle as an indication of valve position is now misleading. In the new position, the lever shows the valve to be fully closed, although the valve is actually in the fully open position. Going by the position of the valve handle, the operators assumed the valve was closed when it was actually open. The operators did not check the valve position provided by the valve manufacturer. See Figure 11.31 for a photograph showing the valve indicator position provided by the valve manufacturer. Figure 11.32 shows the temporary hose that was connected by the operators to depressurize and drain the pump.

US Chemical Safety and Hazard Investigation Board (CSB). Public Domain.

Figure 11.30 Illustration of the suction valve on the alkylation unit alkylate product pump. Note the position of the valve lever or bar, which is perpendicular to the product flow. The operators interpreted this to indicate that the valve was closed, although it was in the fully open position.

US Chemical Safety and Hazard Investigation Board (CSB). Public Domain.

Figure 11.31 Photograph of a rotating plug quarter-turn valve showing the valve position indicator provided by the valve manufacturer. Note that the position indicator shows this valve in the fully opened position.

US Chemical Safety and Hazard Investigation Board (CSB). Public Domain.

Figure 11.32 Photograph of the alkylation unit alkylate product pump with a temporary hose attached, used to drain and depressurize the pump. Note that the pump case drain is also shown, but was not used to verify the pump's zero energy, as it was not equipped with a valve to isolate it from the line used for depressuring the pump.

As I mentioned earlier, the operators also connected a temporary hose to drain and depressurize the pump before handing it over to the maintenance crew. This was a good effort on their part; however, it turns out that either the bleeder or, most likely, the hose was obstructed, and the pump remained fully pressurized. Using the hose did not give the operators an independent way to confirm that the pump was isolated and depressurized.

During the CSB's investigation of this incident, they discovered that one of the pump mechanics had noticed that the valve position indicator on the suction valve body showed that the valve was open; however, he did not relate this information to his co-workers.

The CSB also reported that the plant operator placed tags and locks on the suction and discharge valves to prevent inadvertent opening and indicate that they had been closed. The mechanic specialist then placed tags and locks on the suction and discharge valves. When the mechanic returned, the mechanic specialist told him that the valves had been closed, secured, tagged, and locked per the facility's LOTO procedure and that they could remove the pump. When the temporary hose was removed, naphtha drained for a couple of minutes and quickly subsided, indicating to the workers that the pump was isolated, drained, and depressurized.

As the worker opened the pump case flange, alkylate was suddenly released at about 150 pounds per square inch gauge (psig) and 350 °F. The release produced a loud roaring sound that could be heard throughout the refinery.

The mechanic was blown over by an adjacent pump and suffered broken ribs. Material was blown into the mechanical specialist's eyes, so he went to an eyewash station, cleared his eyes, and quickly exited the unit. Alkylate, which covered the plant operator's clothing, quickly ignited, seriously burning the operator in the ensuing fire. About 30–45 seconds after the initial release, the first of several explosions occurred.

Author's Comments

I noticed this comment in the CSB report: "Before leaving the area to obtain materials needed to remove the pump, the mechanic noticed that the valve position indicator on the suction valve body showed that the valve was open; he did not relate this information to his co-workers." In a later reference, the CSB stated that the same mechanic then started the pump removal by working on flanges on the opposite side of this valve position indicator. At that point, he observed that the valve handle was in the perpendicular position and believed the valve had been closed by the operator. The point is that the mechanic noticed the valve position indicator, which showed the valve to be fully open, but did not bring this to the operator's attention.

The point I make is that if the mechanic had told the operator when he first noticed the valve indicator position, this incident might not have happened. My experience is that mechanics are quite sharp and can notice details, even if they often don't mention them to the operator. I'm not exactly sure why; they might have full confidence in the operator's ability to handle things correctly, or they could be somewhat intimidated by the operator's strong knowledge of the unit and its operation.

Whatever this may be, I believe it suggests that we need to improve our communication with the mechanics on our process units and make it clear that whenever they see something that doesn't look right, they should speak up immediately. This is especially important during the start-of-job meeting or briefing. It's the ideal time to discuss this with everyone involved in the task. Take a couple of minutes to have this conversation with all involved, and feel free to use this case study as an example of what can happen.

Another important point I believe should be highlighted is the strict adherence to the MOC procedure. When the valve was modified at the outside shop, this change should have prompted an MOC document and approval process prior to implementation. I understand that this may have appeared to be a very minor change; however, we can see the direct outcome resulting from it. We must always follow the strict requirements of the MOC process, whether the changes are made on the unit, in our shop, in an outside shop, or in an engineer's work area. All modifications should be thoroughly evaluated for their potential safety implications.

Incident 22: Loss of Containment Due to the Failure to Achieve Energy Isolation

Equilon Enterprises LLC. Wilmington, California

July 31, 2000; zero fatalities, one serious injury (10 days hospitalized)

While observing a steam cleaning operation, the night supervisor fell from a 21-foot ladder onto an asphalt surface. He received serious fall injuries and was hospitalized for ten days. OSHA found that the employee was exposed to hydrocarbon vapors and hydrogen sulfide before falling.

The OSHA prepared the following report.

At approximately 10 p.m. on July 31, 2000, employee #1, the night supervisor, was observing the progress of a steam cleaning operation involving the out-of-service vapor recovery line to tank #1025. He was investigating the area around the 10-inch blind flange and the spreader bars, which were reportedly plugged, when a stream of water began to flow from the top of the flange.

Employee #1 began to smell hydrogen sulfide gas and saw a stream of water and light hydrocarbons start to gush out of the area around the spreader bars and cover the 6 feet square scaffold platform. He later stated that he held his breath, climbed over the railing, and began to descend the 21 feet scaffold ladder. Shortly thereafter, his hands and feet slipped from the rungs, and he fell to the asphalt floor. Employee #1 suffered a fractured right femur, right knee, and left shoulder, and was hospitalized for 10 days at Long Beach Memorial Hospital.

Among the causal factors that led to the accident were the lack of controls to safely dispose of the oil-water mixtures and the foreseeable toxic effects of hydrogen sulfide during maintenance activities. There were no bleed valves on the system to relieve pressure in the piping, and no covers/diverters over the flanges to direct oily water down instead of onto the ladder rungs.

https://www.osha.gov/ords/imis/accidentsearch.accident_detail?id=201033305

Author's Comments

- Removing nonessential people from the work area when opening equipment
- Emphasize the use of warning signs and barricades in areas where equipment is being opened
- The criticality of workers wearing hydrogen sulfide detectors

While this supervisor was seriously injured, he was fortunate to have survived. This incident highlights why we should strive to eliminate nonessential people from work areas where OPE operations are in progress. This includes activities such as purging equipment or clearing or cleaning operations to prepare for OPE. Many hazards may be present during these operations, such as hot steam or condensate, toxics such as hydrogen sulfide (H_2S), asphyxiants such as nitrogen, or other similar hazards are not uncommon.

The use of hydrogen sulfide detectors and training for employees and others to quickly leave the area when the detector alarms can and will save lives. I believe that hydrogen sulfide detectors are invaluable pieces of equipment and that everyone who is working on a process unit that contains, or processes high concentrations of hydrogen sulfide, should be wearing a detection device each and every time they venture onto the unit or area where H_2S is present.

This also speaks to the effective use of barricades, warning signs, and other devices to protect personnel and ensure that those who are not directly involved are not entering the work area during these operations.

Additional Reference

Occupational Safety and Health Administration (OSHA), Employee Injured in Fall from Scaffold Ladder, https://www.osha.gov/ords/imis/accidentsearch.accident_detail?id=201033305.

Incident 23: Loss of Containment Due to the Failure to Achieve Energy Isolation

Valero Energy Corporation, Corpus Christi, Texas

November 29, 2000; one fatality, one injury

Personnel exposure to hydrogen sulfide while removing an isolation blind on a sulfur plant (working without protection by breathing air). The source of H_2S was from a leaking 12″ overhead gas valve that was closed and locked out, but was leaking internally.

The following report was prepared by the OSHA.

Employees #1 and #2, who worked for a nested maintenance contractor, were finishing the turnaround of the sulfur recovery complex at the Valero Energy Refinery in Corpus Christi, Texas. They were removing a 14-inch isolation blind from the overhead inlet of a horizontal receiver vessel. The vessel was part of an amine treating unit that had been emptied, steamed out, and drained a few days before.

After several attempts, the overhead piping had been replaced, and the blinds had been removed and reinstalled. Employees #1 and #2 were working from a scaffold when they were exposed to strong hydrogen sulfide emissions from the vessel. Employee #1 staggered away, but within minutes had lapsed into unconsciousness and died. Employee #2 managed to escape and reach grade level. He was hospitalized for observation and released with no lasting effects.

The vessel had accumulated sour gas from a connected overhead gas line, tied to nearby sulfur trains that were operating at relatively low pressure. The source was a single leaking 12-inch gate valve had been closed and locked out. Employees #1 and #2 were working without respiratory protection or gas detection equipment. The valve inspection program, LOTO program, and respiratory protection were found to be lacking. At the time of the accident, the foreman was also overseeing other crews at the site.

Author's Comments

- Awareness of areas containing super-high hydrogen sulfide concentrations
- Tracking of isolation blinds and blanks
- Importance of workers' use of H_2S detectors and SCBA
- Importance of equipment walkdowns before issuing work permits

Unfortunately, this incident has some common factors with a similar incident at this same site, which occurred in December 1998. See Incident 4 above or the OSHA website for more details.

Petroleum refinery sulfur plants are designed to convert superconcentrations of hydrogen sulfide to elemental sulfur. These high-concentration H_2S streams are typically directly from Amine scrubbers on other process units such as diesel and NHTs and sour water strippers, and the feed gas is extremely high in H_2S concentration. It is common for these streams to be at 80–90% hydrogen sulfide levels or even higher. This includes streams and process vessels associated with sulfur plants and the pipelines from the amine scrubbers to the sulfur plants.

This means that streams to the sulfur plants will have concentrations of H_2S at 800,000–900,000 ppm or higher. When you consider that the IDLH for hydrogen sulfide is 100 ppm and that a single exposure to concentrations between 1,000 and 2,000 ppm can result in instantaneous death. It only takes one breath of H_2S at these concentrations to kill a human, and deaths have occurred at much lower concentrations. People have died due to exposures of about 500 ppm.

This is why I prefer that all workers in these areas wear H_2S detectors at all times. Additionally, I believe all workers working on or with these highly concentrated H_2S systems should utilize SCBA.

It is essential for each facility to have a detailed tracking program for all isolation blinds. Each blind should be tracked by its location and number, entered into the database and associated drawing when it is installed in the flange, and then removed from the list when the blind is removed. The blind list should always present a current view of the installed blinds and must remain up to date.

Due to the high concentration of H_2S, this work is serious. Simply being near these systems without full precautions can put all personnel at risk. All maintenance activities or tasks where an individual works against an unblinded valve or connection should be performed while using either hose line breathing equipment or, at a minimum, a SCBA.

Additional Reference

US Occupational Safety and Health Administration (OSHA), Accident Summary Nr: 200210755 – One Killed, One Injured After Inhaling Hydrogen Sulfide, https://www.osha.gov/ords/imis/accidentsearch.accident_detail?id=200210755.

Incident 24: Loss of Containment Due to the Failure to Achieve Energy Isolation

Tosco Avon Refinery, Martinez, California

A major unit fire occurred on February 23, 1999, while workers were in the process of removing a section of light naphtha overhead piping for replacement on an operating crude atmospheric fractionator.

As a result of the workers being trapped in the structure, four workers were killed, and one was seriously burned (four fatalities and one serious injury).

On February 23, 1999, four people were killed, and one was critically injured during maintenance work to remove and replace a corroded naphtha line on an operating crude unit. The CSB investigated this incident, and a detailed report was provided with their investigation results. I encourage the readers to access the final CSB report using the link in the "Additional References" section of this chapter. A summary of this report is provided here.

This CSB report is well written, and as a result, I have learned many great process safety lessons. For almost twenty years, I have used material from this report as a class exercise in my Safe Operations Training course. There is always a great reception and feedback from this exercise.

The Tosco Martinez Refinery was a relatively small, single-train refinery with a single operating crude unit. On the day of the incident, the crude unit was operating; however, the atmospheric overhead naphtha piping was experiencing leaks due to severe internal corrosion, which was attributed to poor performance on the crude desalter. The desalter appeared to be undersized for the crude oil quality and feed rates being run. The desalter was routinely operated above the design parameters, and there was excessive water in the crude feedstock. The crude feed was also significantly heavier than the original design. The result is that excess water and salt were being sent to the atmospheric column, resulting in severe corrosion in the atmospheric column and the overhead system. The CSB determined that for the year before the incident, the desalter had run about 40% above design, significantly decreasing the efficiency of the desalter operation.

On February 10, about two weeks before the deadly incident, a pinhole leak was discovered on the inside of the top elbow of the 6″ (15 cm) naphtha draw piping, near where it was attached to the upper section of the atmospheric tower at 112′ (34 m) above grade. Attempts were made to isolate the piping by closing four valves (one at the tower and three at the naphtha control valve station located at grade). The unit remained in operation, and inspection subsequently determined that the pipe was corroded and required replacement.

The small leak reoccurred several times during the period while inspection was evaluating the pipe and while preparations were being made for its replacement. The operations crew had been attempting to isolate the naphtha overhead piping circuit; however, they were hampered by leaking isolation valves due to internal plugging and corrosion. The operations crew attempted to isolate the circuit for 13 days without success. Reportedly, management scheduled a maintenance crew to replace the severely corroded section of the line while the unit was still in operation, even though operations were still working to isolate the system. Figure 11.33 from the CSB final report shows the crude fractionator and the naphtha stripper with the interconnecting piping.

Several attempts were made to isolate the piping; however, the valves were leaking through, and the leak continued to recur even though the valves were tightened and retightened. On February 23, a work permit was issued to "drain and remove the piping." Attempts to drain the piping through bleeders near the bottom of the line at the control valve station failed as the bleeders were plugged. Following unsuccessful attempts to clear the bleeders, a maintenance supervisor directed the workers to make two cuts into the piping with a pneumatic saw. The first cut was made successfully at about 104 feet (31 m) above grade. However, while cutting the line at a location about 78 feet (23 m) above grade, naphtha started leaking from the saw cut. The repair operations ceased, and the second cut was closed with a line clamp to stop the leak. The maintenance supervisor then directed the workers to open a flange at 38 feet (12 m) above grade to drain the line by directing the naphtha onto plastic sheeting and into a drain pan where a vacuum truck was waiting to pick up the drained naphtha.

When the flange was opened, and while the line was draining, naphtha was released from the open cut at the 104 feet (32 m) elevation, resulting in naphtha spilling onto the scaffold and the surrounding area. The spilled naphtha immediately flashed into a large fire engulfing the scaffolding and the workers. The workers' actions are illustrated in Figure 11.34.

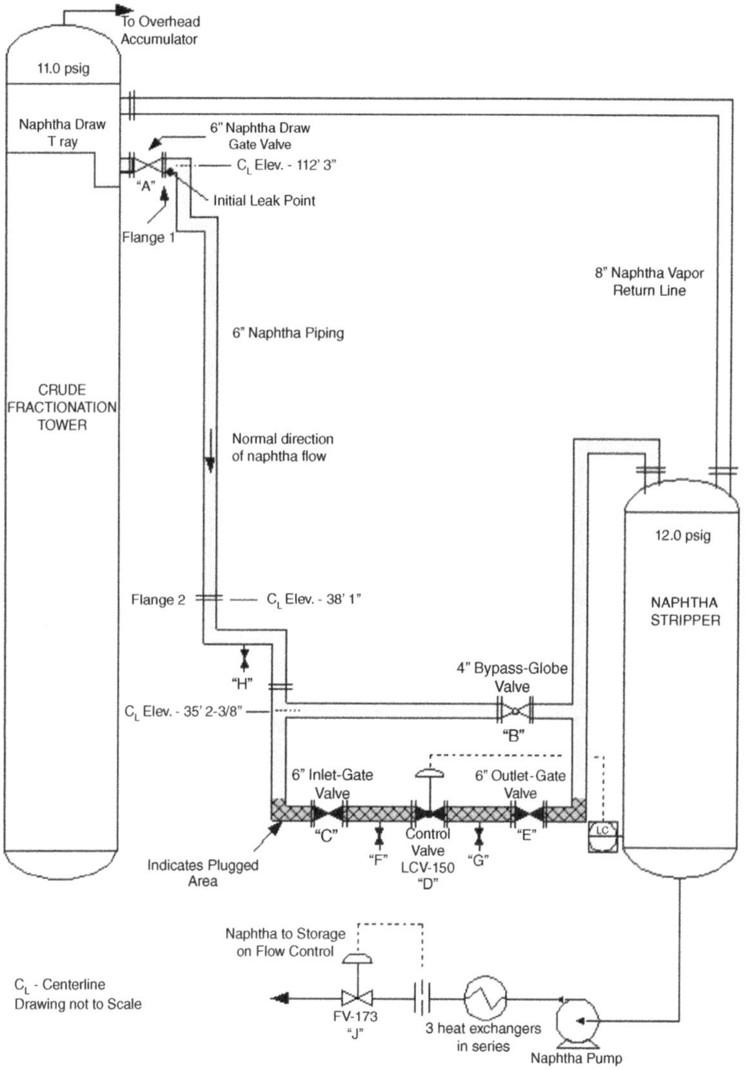

Figure 11.33 Crude unit atmospheric fractionator and naphtha draw, simplified flow diagram at the time of the initial leak. US Chemical Safety and Hazard Investigation Board (CSB). Public Domain.

One worker was pronounced dead at the site, and three workers subsequently died at the hospital. The fifth worker was able to jump from the scaffold but received critical injuries as a result of the incident.

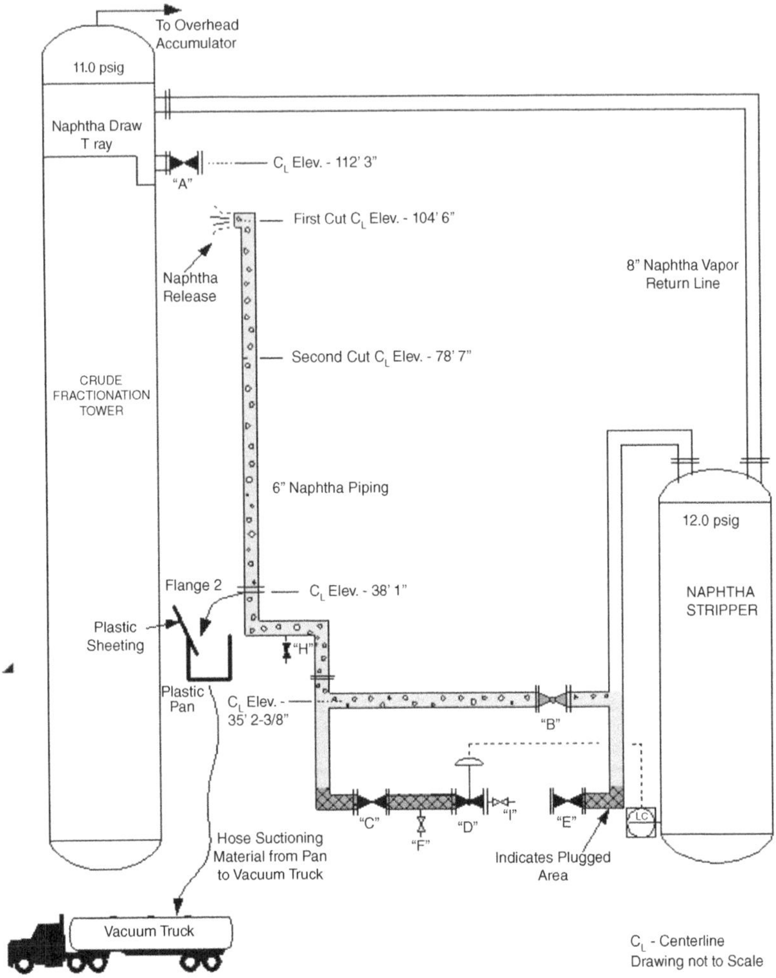

Figure 11.34 Crude unit atmospheric fractionator and naphtha draw, simplified flow diagram at the time of the large naphtha release and fire. US Chemical Safety and Hazard Investigation Board (CSB). Public Domain.

Other Background Information

- The refinery has two crude units. One of the units was shut down and was being decommissioned at the time when this incident occurred.
- The unit operator issued the work permit to "drain and remove the piping." The permit was considered as "nonhazardous" work since it did not involve hot work or confined space entry (as per the refinery work rules).
- The maintenance supervisor was directing the work on-site.

- Consistent with other "nonhazardous" work, there was no process supervision or safety representatives overseeing the work activities to replace the naphtha piping.
- The job planning process did not require a formal evaluation of the hazards of replacing the naphtha piping since the task was considered "nonhazardous" work.
- The work permits issued to drain the piping did not cover precautions for a benzene-containing stream or the use of a crane, vacuum truck, or pneumatic saw. The work permit specified drain and removal as the work to be completed.
- The valves were observed to be leaking internally several times between February 10 and February 23. The stripper tower also filled several times during this period and was pumped to storage to control the level.
- The refinery was operating the crude unit desalter beyond its design parameters, allowing the potential for water, salts, and other contaminants to enter the process unit with the crude oil. The desalter was reportedly being operated at about 40% above design limits.
- Attempts to drain the piping from bleeders at the control valve station were not successful. The isolation valves, piping, and drains located at the naphtha control valve station were found to be plugged with solids.

  Note that Figure 11.35 shows the isolation valve stem for the control valve station was found several inches open, indicating that the valve was not fully closed or was obstructed. Figure 11.36 clearly shows the condition of the isolation valve during the CSB investigation.
- The most recent documented PSM safety audit was completed during 1995 (required by Cal/OSHA every three years). No documented audits of line breaking, LOTO, or blinding procedures or practices were done in the previous three years.
- Workers had been informed by refinery management that they have the authority to stop work if it is deemed unsafe.

Key Lessons Learned

Lack of Effective Job Planning and Evaluation of Hazards

CSB Board Member Irv Rosenthal reported in a CSB News Release, "Better management of job planning and execution could have prevented this tragedy. Our investigation uncovered two root causes of the accident. First, Avon Refinery management did not have an effective process for assessing the dangers of maintenance operations and implementing needed safeguards. Second, neither the parent Tosco Corporation nor the facility management had investigated or corrected a pattern of serious unsafe practices at the Avon Refinery."

US Chemical Safety and Hazard Investigation Board (CSB). Public Domain.

Figure 11.35 A close-up photograph of the valve stem of block valve (C), upstream of LCV-150, with the valve wheel fully tightened. The protruding stem shows the valve to be jammed partway open, indicating possible material blockage in the line.

The CSB found that the job planning did not require an evaluation of the hazards of replacing the naphtha overhead piping with the unit remaining in operation. Compounding this, the piping repair work was classified as low-risk maintenance, even though the operators were having trouble isolating the piping system from the running unit. The CSB said in their report that "despite serious hazards caused by the inability to drain and isolate the line – known to supervisors and workers during the week prior to the incident – the low-risk classification was not reevaluated, nor did management formulate a plan to control the known hazards."

The work was being carried out with the unit in service (running), and operations were unable to effectively isolate the line being removed from the operating unit.

Effective Maintenance Procedures

Mr. Rosenthal also reported that "The refinery's own procedures directed that the piping should be isolated and drained prior to attempting this kind of repair. This procedure was not followed. Opening a pipe containing naphtha in the presence of multiple ignition sources was a formula for disaster. The entire process unit

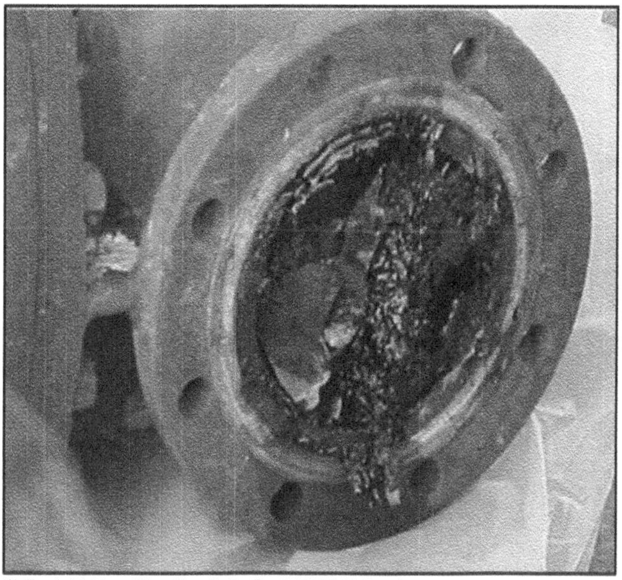

US Chemical Safety and Hazard Investigation Board (CSB). Public Domain.

Figure 11.36 A close-up photograph of the material blockage of block valve (C), upstream of LCV-150.

should have been shut down, which would have eliminated ignition sources and allowed the naphtha to be fully and safely drained."

Operations Involvement and Oversight

Key decisions, including when and how to cut the piping, were left to maintenance personnel. Operators were not involved and were not on site when these decisions were made and carried out.

Management Involvement and Oversight

"Management had a responsibility to see that work was halted, and should not, as has been suggested, have relied solely upon individual workers to stop an unsafe activity. Worker error was not the root cause of this accident. A satisfactory management system is one that, anticipating that humans will inevitably make mistakes, still ensures the safe conduct of work. Effective job planning, hazard review, and management oversight could have prevented the Avon tragedy."

See Figure 11.37 for an image of where this tragic accident happened and the fire damage to the fractionation column.

After reviewing this incident, you may wonder why I included it in this book, since it happened a long time ago and our industry standards and expectations have improved significantly since 1999. I fully agree that our internal standards,

US Chemical Safety and Hazard Investigation Board (CSB). Public Domain.

Figure 11.37 Photograph of the fractionation column and the fire damage.

industry regulations, and company and community expectations have changed dramatically during this period and have markedly improved. Unfortunately, the other point that needs to be made is that these incidents can and do still happen today, and workers are still gravely injured or killed. Incidents continue to happen most often because we fail to follow our internal procedures, standards, and regulations or take shortcuts.

Whether you are a supervisor, manager, engineer, or operator, we must set standards of excellence when executing operational tasks in the field. This means that procedures and standards must be respected and followed. Critical procedures, procedures that can result in loss of life or serious injuries or loss of containment of flammable or toxic materials, must be closely followed. Critical procedures should be developed on a step-by-step basis with operator's initials required after the execution of each step. This ensures that critical steps are not missed or omitted and that steps are executed in the correct sequence.

In this incident, the workers never isolated the equipment. The equipment was opened to the atmosphere without properly draining or isolating using blinds or other effective isolation. This deviation from established procedures was a recipe for disaster. Of course, there were other significant procedural issues and a lack of proper oversight and management.

If a deviation from a procedure is necessary, or if there is no procedure for the task, we must complete a deviation request using the MOC process and obtain approval from at least one level up. This is typically achieved by redlining the existing procedure with the proposed changes and securing approval for the red lines from the manager at least one level above.

Effective procedures require significant time and resources to develop. They are designed to account for nearly all contingencies and include mitigations to ensure safe execution. However, if procedures are not followed as intended or if workers take shortcuts, they become ineffective.

Additional References

Digital Refining, Multiple Corrosion Mechanisms in Crude Distillation Overhead System, https://www.digitalrefining.com/article/1000598/multiple-corrosion-mechanisms-in-crude-distillation-overhead-system.

Digital Refining, Real-time Crude and Desalter Monitoring, https://www.digitalrefining.com/article/1002501/real-time-crude-and-desalter-monitoring.

Slideshare, Assignment 1 – Crude Oil Desalting Unit.pdf, https://www.slideshare.net/slideshow/assignment-1-crude-oil-desalting-unitpdf/264202784.

Sulzer, Crude Oil Desalting – Reliable Solutions for Critical Services, https://www.sulzer.com/en/shared/applications/refining-crude-oil-desalting-process.

US Chemical Safety and Hazard Investigation Board (CSB), Tosco Avon Refinery Petroleum Naphtha Fire, https://www.csb.gov/tosco-avon-refinery-petroleum-naphtha-fire/.

Dr. Maurice J. Wilkins C.Eng, FIChemE, FInstMC, ISA Fellow, Following Procedures Can Save Lives! https://refiningcommunity.com/wp-content/uploads/2015/05/Following-Procedures-Can-Save-Lives-Wilkins-Yokogawa-DCU-Galveston-2015.pdf.

Incident 25: Loss of Containment Due to the Failure to Achieve Energy Isolation

Citgo Refinery, Corpus Christi, Texas

December 14, 1998; Zero fatalities, ten were transported to the hospital, five hospitalized (This incident involved an attempted rescue)

Hydrogen Sulfide was released from an open connection on a new 6-inch exchanger bypass line. H_2S penetrated near and inside the control room and through nearby process units.

The following report was prepared by the OSHA.

Employees #1 through #10 worked at a crude oil refinery (Citgo Refinery, SRU West Plant, Corpus Christi, Texas 78469) at which nearby process units (crude oil and gas oil treater) were undergoing a planned turnaround. The units had staggered shutdown and startup schedules, and they shared some product transfer piping leading to storage.

When the product flow was rerouted by operations prematurely, sour naphtha liquid was released into the crude unit from an open connection installed on a new 6-inch piping section that served as an exchanger bypass line. The low-pressure release lasted for several minutes and allowed clouds of toxic hydrogen sulfide gas to accumulate in and around the nearby control room and through nearby process units.

Emergency alarm and evacuation procedures were not executed according to plan. Operators responding in the field at first attempted rescue efforts without using emergency air packs, ultimately adding to the total of exposed employees. Numerous exposures to the evolved sour gas resulted in injury to both plant and contractor employees. Employees #1 through #10 were transported to the hospital, and Employees #2 through #4, #9, and #10 were admitted. None experienced permanent effects.

Subsequent investigation revealed weaknesses and discrepancies in the plant LOTO program, and poor implementation and upkeep of the process safety program. Unit isolation blinds and LOTO blinds for the equipment were not clear to the operators due to poor tracking.

Operating procedures were both outdated (e.g., the use of checklists) and/or unclear, such as how to perform the final stages of unit startup.

Author's Comments

- Requirements for detailed equipment walkdown before commissioning
- Criticality of the isolation blind tracking
- The value of H_2S detectors and the importance of wearing them when working near H_2S-containing equipment

When commissioning new or modified equipment, we should complete a detailed walkdown to review that all connections and tie-ins are properly secured and managed. This includes all new or modified piping to ensure all connections are secure with proper gaskets and bolted flanges. Flange bolts or studs should be verified as correct and tightened or torqued if necessary.

When installed, all blinds should be listed in a detailed blind inventory. The blind should be added to the list immediately upon installation and removed from the list when taken out of the unit. The blind inventory should always accurately reflect the current status of the unit's blinds.

SRU units, also known as sulfur plants, process highly concentrated hydrogen sulfide. Feed gas typically contains 80–90% H_2S, or 800,000–900,000 ppm. The IDLH is only 100 ppm. On units with significant hydrogen sulfide levels, all personnel must wear an H_2S detector, positioned as close to the breathing zone as possible – not on the hip or back pocket. H_2S is a deceptive gas because it deadens olfactory nerves at around 100–150 ppm, making it undetectable by smell. People have died at concentrations below 500 ppm, and at about 1,000–2,000 ppm, death can occur almost instantly. One breath of SRU feed gas can result in immediate death.

Additional Reference

US Occupational Safety and Health Administration (OSHA), Ten Workers Were Exposed to Hydrogen Sulfide Release on December 14, 1998. There are two different reports, one for Coastal Refining and one for Citgo Refining (both in Corpus Christi, Texas), https://www.osha.gov/ords/imis/accidentsearch.accident_detail?id=200210391; https://www.osha.gov/ords/imis/establishment.inspection_detail?id=302099908.

Incident 26: Loss of Containment Due to the Failure to Achieve Energy Isolation

Phillips Polyethylene Plant, Pasadena, Texas

This incident occurred due to the loss of containment of ethylene and isobutane, resulting in the massive explosion and fire, which destroyed the facility. The property damage was estimated to be approximately $715 million or about $1.5 billion in today's dollars, plus additional business losses estimated to be equivalent to the property damage. Figure 11.38 is from the investigation report by the U.S. Federal Emergency Management Agency and shows the Phillips Polyethylene facility before the devastating 1989 explosion.

October 29, 1989; twenty-three fatalities, three hundred and fourteen others injured, many seriously

The Phillips polyethylene plant produced about 15 billion pounds (6.8 million tons) of high-density polyethylene (HDPE) annually, a plastic used in making milk bottles and other containers. The Houston Chemical Complex (HCC) employed 905 company workers and around 600 daily contract workers involved in routine maintenance and new plant construction.

The Phillips polyethylene process uses a slurry method with a chromium oxide catalyst supported on silica gel, known as the Phillips catalyst. It is prepared and activated at high temperature (calcined) before being injected into the reactors. The catalyst is dissolved in a solvent like isobutane and circulated through large continuous loop reactors, typically several hundred feet tall. The reaction takes place at elevated temperature and pressure, producing solid polyethylene particles called "fluff." Hydrogen is usually added to the process to control the molecular weight of the polyethylene.

The reactors are designed with tall vertical piping loops with settling legs at the bottom of each loop. These settling legs collect the plastic powder, which appears as white dust and is referred to as "fluff," formed by reacting the ethylene in the reactors. The fluff is removed from the settling legs, residual gas is separated, and then the powder is fed to extruders to produce plastic pellets. These pellets are

FEMA. Public Domain.

Figure 11.38 Aerial view of the Phillips plant prior to the 1989 explosion.
Note: Portions of this incident review were used with permission of the publisher (Wiley) from Darryl's first book ("Process Operations Safety, The What, Why, and How of Safe Petrochemical Plant Operations").

stored in silos, segregated by grade, then blended to the customer's specifications and shipped in rail cars to consumers worldwide.

Each reactor loop has six settling legs used to control reactor pressure and discharge products to the onsite holding silos. Near the bottom of each settling leg is a pneumatically operated ball valve, called by the manufacturer's name, the Demco valve. The Demco valve acts as a dump valve to control pressure and a block or isolation valve when removing the settling leg from service. For example, when the settling leg plugs with plastic powder, the powder forms a "log" of plastic in the settling leg.

Unfortunately, the design did not include an effective way to isolate the settling legs from the reactors. There were no double block valves or bleed arrangements, as mentioned in other case studies. The method being used was to close the Demco valve and remove the pneumatic lines from the actuator inlet and outlet connections to prevent the valve from being inadvertently opened while maintenance was underway on the settling legs. See Figure 11.39 for an illustration of the reactor settling leg and the Demco valve.

On the day of the incident, several settling legs were plugged with packed fluff, and an operation was underway to open them and remove the plastic material. The piping for the settling leg was isolated from the active reactor using the Demco valve for isolation. Additionally, the air supply lines to the pneumatic

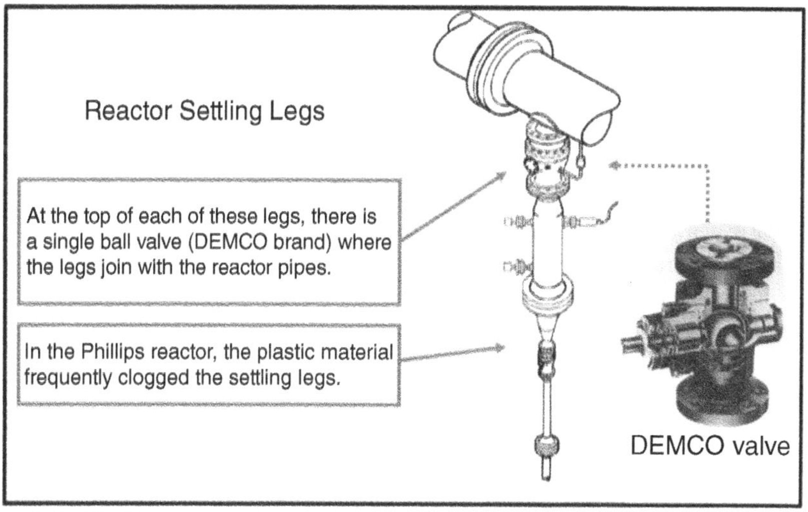

FEMA. Public Domain.

Figure 11.39 Reactor Settling Legs – Located at the bottom of the polyethylene loop reactors.

valve actuator were removed to prevent the valve from opening unexpectedly. The piping was disassembled during maintenance to facilitate unplugging.

We don't know what exactly happened in this event since everyone in this work team was killed in the resulting explosion. What we do know is that a significant release of flammable vapor occurred while routine work was being performed to remove accumulated polymer from an HDPE reactor settling leg. We also know that the settling leg had not been reassembled, and parts of the pipe remained disconnected, indicating that the work was still in progress when the incident occurred.

Most importantly, the ball valve, the Demco valve, was found fully open even though the maintenance work was not finished and the piping had not been reconnected. The pneumatic actuator lines to the ball valve had also been reconnected, but, unfortunately, reversed, which allowed the valve to open completely with the settling leg piping not fully connected. This caused the pressurized product to be released directly into the atmosphere, resulting in a significant loss of containment of the highly flammable product and a massive explosion.

At the time of the incident, the reactor operated normally at about 700 psig and contained a highly flammable mix of Ethylene and Isobutane. The released vapor was quickly ignited, causing the vapor cloud explosion (VCE). This event caused 23 deaths and 314 injuries and destroyed two of the site's HDPE production plants. The explosion was felt miles away from the facility and registered between 3 and 4 on the Richter scale on Rice University's seismographs, about 25 miles from the site. There were also several secondary explosions, including the boiling liquid expanding vapor explosion (BLEVE) of other on-site pressurized storage tanks.

The incident also underscores the importance of strict worker adherence to energy isolation procedures. Other contributing factors may include the timing of Process Hazard Analysis and a more rigorous work permit process for OPE. Additionally, key lessons involve the location of occupied buildings near the process units, the need for better fire protection equipment, and improved personnel training in work permits, maintenance procedures, and emergency/evacuation protocols. See Figures 11.40 and 11.41 for photos of this catastrophic event. An image showing the area affected by this massive explosion is displayed in Figure 11.42. Note that debris from the explosion was found 6 miles from the facility!

Author's Comments

The explosion was heard for miles in all directions; it registered a 3.5 on the Richter scale at Rice University, 25 miles from the site, and debris was found 6 miles away. It has been estimated that this would have been equivalent to about a four on the Richter scale at ground zero. Federal labor officials claimed the initial explosion had a force of 2.4 tons of TNT. Several additional explosions occurred after the

FEMA. Public Domain.

Figure 11.40 Aerial view of the Phillips Polyethylene facility after the explosion. The photo is seen from east to west. The arrow highlights a berm to the west of the property, which is a bank for a large holding pond that served well as a protective berm.

FEMA. Public Domain.

Figure 11.41 Aerial view of the Phillips Polyethylene facility after the explosion. View of the plant from west to east. This view has the south side to the right.

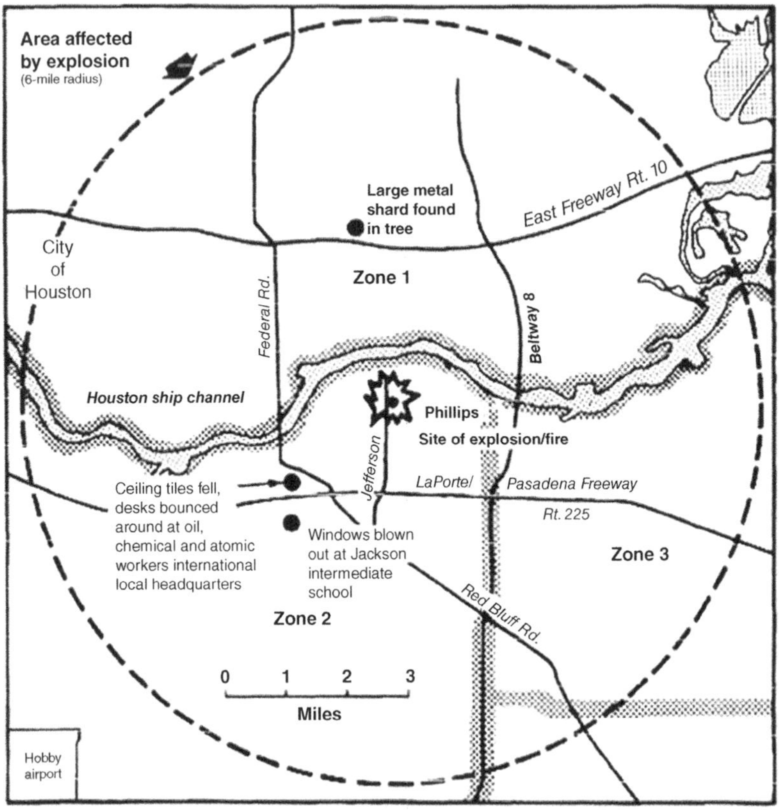

Figure 11.42 View of the area affected by the Phillips Polyethylene explosion. Note that the explosion registered 3.5 on the Richter scale at Rice University (approx. 25 miles from the facility); some reports indicate that it was estimated to be 4.0 at the epicenter. Metal objects were found 6 miles from the facility. FEMA. Public Domain.

main blast, including several process vessels that failed due to BLEVE. This was a devastating and tragic event with 23 lives lost and many people injured, many of whom suffered life-changing injuries. As a result of this incident, numerous lives were forever changed. This is what process safety is about, and the reason I continue to do what I do today. It is all worthwhile if we can prevent even one of these tragic events or save even one life.

I don't like to speculate about what happened or what the crew was doing when the incident occurred. However, I have heard they were possibly in a hurry to finish the job, and someone tried to reconnect the actuator lines before the piping had been reconnected. Another possibility was that the plastic log was further up in the settling leg, and they couldn't reach it, so they attempted to pressurize the piping to clear it.

In any case, it is clear that the energy isolation was broken. The Demco valve was opened while the reactor was in service and pressurized at about 700 psig with ethylene and isobutane, while the settling leg piping remained open to the atmosphere. What is also clear is that the pneumatic lines to the Demco valve actuator had been reinstalled in reverse order while the settling leg piping was still open.

Lessons from this incident highlight the need for more thorough energy isolation before OPE, such as using double block and bleed isolation valves and employing blinds and blanks to ensure safety from hazardous energy. Another, what appears to be a simple and effective isolation, would be to design incompatible air fittings to prevent reversing the air connections to the pneumatic actuator.

Additional References

American Institute of Chemical Engineers (AICHE), AIChE: 'Living Through a Plant Explosion', https://www.isssource.com/aiche-living-through-a-plant-explosion/.

British Health and Safety Executive (HSE), Phillips 66, Pasadena, USA. October 23, 1989, https://www.hse.gov.uk/comah/sragtech/casepasadena89.htm.

Federal Emergency Management Agency – United States Fire Administration, Technical Report Series – Phillips Petroleum Chemical Plant Explosion and Fire, Pasadena, Texas, https://ncsp.tamu.edu/reports/USFA/pasadena.pdf.

Institution of Chemical Engineers (IchemE), Explosion at the Phillips' Houston Chemical Complex, Pasadena, October 23, 1989, Loss Prevention Bulletin 272, April 2020, https://www.icheme.org/media/13605/lpb272_pg29.pdf.

Plastics News, The Day that Rocked an Industry, https://www.plasticsnews.com/article/19990308/NEWS/303089943/the-day-that-rocked-an-industry.

Purdue University, 30 Years Later: Remembering the Explosion at Pasadena, https://engineering.purdue.edu/P2SAC/presentations/documents/P2SAC-5-Dec-Pasadena-Vaughen.pdf.

US Occupational Safety and Health Administration (OSHA), Violation Detail – Inspection No.: 106612443, Standard Cited: 5A0001 OSH Act General Duty Paragraph, https://www.osha.gov/ords/imis/establishment.violation_detail?id=106612443&citation_id=01001.

Incident 27: Loss of Containment Due to the Failure to Achieve Energy Isolation

BP Grangemouth Refinery, Scotland, situated on the Firth of Forth, on Scotland's east coast

Loss of Containment and a major fire while attempting to replace a faulty flare system isolation valve.

March 13, 1987; two fatalities and two serious injuries

A second incident involving a fire in a crude oil storage tank while cleaning was underway occurred on June 11, 1987, which is covered in Section IV, Incident 13.

Over a brief period in 1987, from March 13 to June 11, there were three process safety incidents that occurred at the BP Grangemouth Refinery in Scotland. The first incident occurred on March 13, 1987, involving a loss of containment and a major fire that occurred during the planned replacement of an isolation valve on the refinery flare line. This accident killed two maintenance contractors and severely injured two others.

The second incident occurred on March 22, 1987, when the Hydrocracker low-pressure separator exploded due to overpressure. This incident had significant injury potential, but fortunately, no one was injured. The third incident occurred on June 11, 1987, when there was a fire and explosion in a crude oil storage tank at the Dalmeny crude terminal, killing a contract employee.

All three of these incidents were investigated by the UK Health and Safety Executive, and a detailed report with recommendations was issued. These incidents resulted in significant fines and penalties imposed by the High Court in Edinburgh on March 21, 1988, when a fine of £250,000 was imposed for the charge relating to the circumstances of the flare line incident and £500,000 for the second charge arising out of the hydrocracker explosion. The circumstances of the fire at the Dalmeny crude terminal were reported to the Procurator Fiscal at Edinburgh, and a Fatal Accident Inquiry was held from March 10 to March 16, 1988.

In this book, we will focus on the two incidents related to OPE, the flare system loss of containment and major fire which occurred on March 13, and the fire which occurred in the crude oil tank which occurred on June 11. A link is provided in the "Additional References" section of this chapter to the HSE report for all three incidents.

BP Grangemouth Flare System Fire (BP Grangemouth)

The flare system incident that occurred in March 1987, at the British Petroleum Grangemouth Refinery in Scotland, was like the 1999 Tosco Avon Refinery in Martinez, California, where four workers were killed, and one was severely injured. The Tosco incident occurred due to loss of containment and a major fire, which occurred during the planned replacement of a crude unit overhead naphtha piping circuit. The similarity with the Grangemouth incident was that the BP crew was attempting to isolate a live flare line to replace a faulty valve when a loss of containment occurred, resulting in a devastating fire and the deaths of two workers and burn injuries to two others.

When the flange on the isolation valve was opened, "gallons and gallons of liquid were coming from the pipe under pressure," creating a large pool fire directly below where the workers were working from a scaffold. The two workers who were trapped on the scaffold died due to their injuries. Figure 11.43 illustrates the severity of the fire that occurred due to ignition of the vapors when liquid hydrocarbons were released when the flare piping system was opened.

The HSE conducted an investigation and found that approximately 15 months before the incident, the flare line isolation (V17) was leaking internally. A decision was made to replace the leaking valve while the FCCU and Number 1 Flare were offline for a turnaround. During this period, gases from the remaining operating units were rerouted to the Number 2 and 3 flares, allowing the V17 valve to be isolated without disrupting ongoing refinery operations.

The senior refinery staff developed a plan for the isolation of the flare system to allow V17 to be isolated and replaced. However, they focused on ensuring that the remaining flare system was not isolated from operating equipment, thereby protecting the remaining operating equipment. They left the specific valve isolation details for the V17 valve to those responsible for the work.

Photo by United Kingdom Health and Safety Executive (HSE).

Figure 11.43 Image of the massive refinery fire which occurred at the BP Grangemouth Refinery while work was underway to remove the isolation valve in the flare system for maintenance.

The two workers were killed after opening flare piping, which they believed was isolated and depressurized. Although the isolation valves were closed, plugging and valve leakage prevented complete isolation of the piping. As the contractors removed the last bolt from the valve flange, the crane increased its lift, and the spacer suddenly sprang upward. In the words of a survivor, "gallons and gallons of liquid were coming from the pipe under pressure" onto the platform and the ground below.

This resulted in a flammable vapor cloud forming from the rapidly spreading pool of light naphtha on the ground directly below the workers. The vapor reached the breathing-air compressor and flashed back around the work area, engulfing the scaffold and the workers. One pipefitter and the crane operator managed to escape, although they suffered severe burns. The rigger and the second pipefitter were on the side of the valve opposite the ladder. Although they succeeded in reaching the ladder side of the scaffold, they were unable to escape and were engulfed in the fire. Unfortunately, they were both killed. The fire was allowed to burn in a controlled manner for nearly two days while the rest of the refinery was shut down and the flare system purged with nitrogen.

Figure 11.44 illustrates the flare line and knockout drum configuration and how liquid had accumulated in the flare line just prior to the release. The drawing shows how liquid should normally drain from the bottom of the flare line into the flare knockout drum. However, due to internal blockage, this drain was obstructed.

The HSE investigation discovered the following causal factors:

- There was no specific plan developed for the isolation, draining, and preparation of the leaking isolation valve. This was left to the workers who were assigned to the task of replacing the faulty valve.
- The work planning focused on the protection of the operating units to ensure the flare system was not isolated from the running units.
- Due to internal condensation that occurred in the uninsulated large-diameter flare lines, a significant amount of light naphtha accumulated in the flare piping. Unfortunately, this was not identified by the workers and was not drained or otherwise removed from the piping system.
- The flare system contained a significant amount of pyrophoric iron scale that was trapped under the isolation valve seats, preventing the isolation valves from properly seating. Most isolation valves were leaking through.
- The flare system was designed with a smaller bore piping system to drain condensate from the large-diameter flare lines into the flare knockout drum. Unfortunately, this drain was blocked by iron scale and debris, allowing liquid condensate (light naphtha) to accumulate in the flare lines.

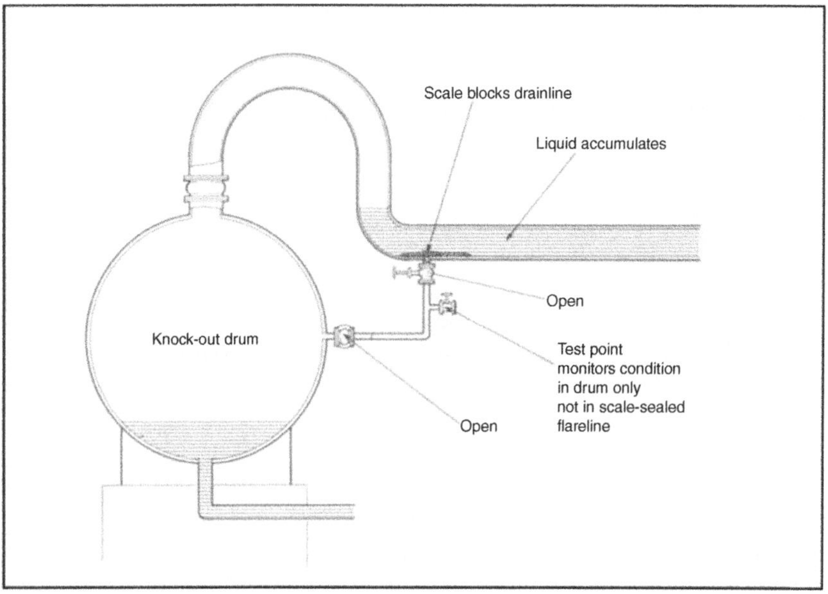

Figure 11.44 An illustration of the large-diameter flare line with the drain from the piping system into the flare knockout drum. Note that how the liquid drain was obstructed by scale, resulting in a significant amount of liquid accumulating in the flare line. Photo by United Kingdom Health and Safety Executive (HSE)/United Kingdom Health and Safety Executive.

- The scaffold platform was not designed with a secondary emergency egress, and egress was partially blocked by the isolation valve.
- The exhaust gas spark arrester was not installed on the diesel-driven breathing-air compressor. There had been no rigorous checks on equipment.

Figure 11.45 is from the British Health and Safety Executive investigation and illustrates how the valve was prevented from closing due to scale trapped under the valve gate. This resulted in the accumulation of liquid in the flare lines contributing to this deadly fire. The investigation report of the HSE of the United Kingdom made the following recommendations:

For senior management (HSE, 1989; pp. 13–14):

- Conduct a detailed hazard analysis during job planning.
- Before delegating work, develop detailed job instructions specific to the particular isolation to ensure the effective draining of flammable liquids.
- Maintain rigorous control over possible ignition sources in the vicinity of maintenance work.

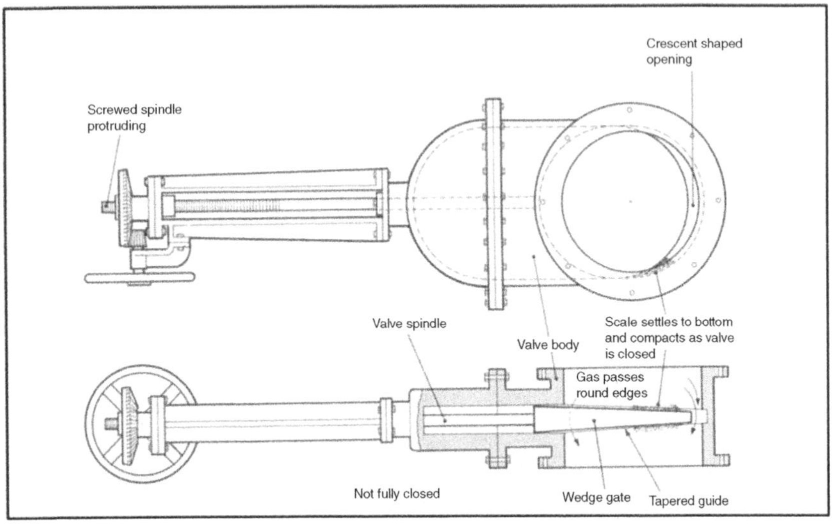

Figure 11.45 An illustration of one of the large-diameter flare isolation valves. Note that how the valve gate was obstructed by scale accumulated in the valve seat and under the gate. The scale prevented the valve from closing, allowing flare gas and liquid to flow under the gate and preventing complete isolation of the flare system. Note that the valve stem protruding from the top of the valve, indicating that the valve was not completely closed. Several of the flare system isolation valves were found to be obstructed in this same way. Photo by United Kingdom Health and Safety Executive (HSE)/United Kingdom Health and Safety Executive.

Author's Comments

Proper planning for safety and verifying zero energy is critical to tasks like this one.

Our role is to ensure that these issues are addressed and that an incident like this does not occur at our facility. Preplanning for a task of this type is essential and should always include planning and verifying the complete isolation of the integrated piping systems. Zero energy should be confirmed before a work permit is issued to open any flanges. If zero energy cannot be confirmed, an alternate plan should be developed, including additional mitigation, and approved by at least one level up in the organization. Under certain circumstances, it may be appropriate to place collection pans to redirect any liquids away from where the people are working. However, we should always strive to demonstrate zero energy.

It is often simple stuff that we miss or fail to get right. Something as simple as providing good and clear emergency egress for the workers can make all the difference when things go wrong in a case like this one. In this case, as outlined in the documentation, there was a single ladder for the four workers and very limited space on the scaffold deck where the valve was to be removed. Reports indicate that the workers working on one side of the valve had to climb over the valve to reach their work location. One ladder and a very limited and congested work area will not serve the workers well in an emergency like this incident.

I also reported in Chapter 4, "When are We Most Vulnerable to These Hazards?" when discussing an incident that occurred at a US Gulf Coast Refinery, where two people died, and several others were critically burned while working to isolate and open a series of stacked reactor feed and effluent exchangers. The situation was similar in that case. Many workers were working from the scaffolds around the exchangers at the time of the release and the resulting fire. Some of these workers were working to burn bolts on the exchanger, while others were in the process of modifying the scaffolding that had been placed around the exchangers. All workers were required by their respective companies to wear fall protection, and all were caught up in the devastating incident when the release occurred and quickly ignited.

Additional Comments by the Author

- There are some great learnings from this incident that should be broadly communicated within the oil and gas industry.
- The work planning process did not address the requirement for isolation of the flare system and the verification of zero energy, but instead left this to the workers. This should be part of the job-site plan, which includes details on completing the isolation and verifying zero energy.
- The work platform and access ladders should be constructed to "green tag" standards, including emergency egress for workers.
- The craftsmen should be trained to "crack" the flange open and verify that there is no liquid or pressure present before fully opening it. Following this procedure, they can quickly reclose the flange if liquid or pressure is detected.
- Liquid draining from the flanges should be a "red flag" warning to all workers, and all work should be immediately stopped. The flanges should be retightened to stop the drainage. The work area must be evacuated right away until the job can be reevaluated, and a new work permit must be issued before work resumes.
- A conductive drain pan should be in place to capture/collect any small residual product that may drain from the piping system when the flanges are opened. The drain pan should be electrically bonded and grounded to prevent ignition from static electricity.
- The work permit should specify that fire protection is required with charged fire nozzles outside the line of fire, but ready to respond if needed.

- The breathing-air compressor should be positioned in an open area well away from the work zone, and the exhaust must be fitted with a spark arrester.
- All other possible ignition sources should be eliminated from the work area.

Additional References

British Health and Safety Executive (HSE), BP Grangemouth 1987 – Incident Summary, https://www.icheme.org/media/13700/the-fires-and-explosion-at-bp-oil-grangemouth-refinery-ltd.pdf

British Health and Safety Executive (HSE), The Fire at BP Oil, Grangemouth Refinery. March 13, 1989, https://www.hse.gov.uk/comah/sragtech/casebpgrange87.htm.

British Health and Safety Executive (HSE), The Hydrocracker Explosion and Fire at BP Oil, Grangemouth Refinery. March 22, 1987, https://www.hse.gov.uk/comah/sragtech/casebpgrang87b.htm.

British Health and Safety Executive (HSE)/Icheme.org/media, A Report of the Investigations by the HSE into the Fires and Explosion at Grangemouth and Dalmeny, Scotland, March 13, March 22, and June 11, 1987, https://www.icheme.org/media/13700/the-fires-and-explosion-at-bp-oil-grangemouth-refinery-ltd.pdf.

Cambridge Safety, Major Accidents – Hazardous Substances, Nebosh Diploma Module C, https://www.cambridgesafety.co.uk/wp-content/uploads/2018/04/Dip-C4-Major-Accidents.pdf.

Incident 28: Loss of Containment Due to Failure to Achieve Energy Isolation

BP Crude Oil Tank Fire at Dalmeny Oil Storage Terminal, Dalmeny, Scotland.

The tank fire occurred while the crude oil storage tank was being cleaned.

June 11, 1987; one fatality

This incident was also investigated by the HSE, and a report was written with recommendations to prevent a recurrence. When I read this report, I could hardly believe the root cause of this tragic incident, although it was presented by the HSE; therefore, I know this to be true.

The Dalmeny oil terminal facility is located about 6 miles from Edenborough and has seven large floating-roof crude oil storage tanks for the storage of stabilized crude oil. Each tank measures 255-foot diameter (78 m) by 61 feet tall (18.5 m).

Tank 807 was emptied, and the floating roof was lowered and rested on 219 support pillars or roof legs. Three roof manhole covers were opened to allow natural ventilation; however, no forced ventilation was provided to help prevent

United Kingdom Health and Safety Executive (HSE).

Figure 11.46 Photograph of Tank 807 at the Dalmeny Oil Storage Terminal. This photograph was taken as workers searched for the victim of an internal fire in the storage tank. The fire occurred as one of the workers was smoking inside the tank while cleaning sludge and residual crude oil from the tank.

the accumulation of hydrocarbon vapors within the tank. Three workers were working inside the storage tank, removing residual oil, and a fourth worker was stationed outside the tank along with a BP employee. The three workers inside the storage tank were wearing breathing apparatus supplied by a breathing-air compressor located outside the tank bund (tank firewall). Figure 11.46 was included in the British Health and Safety Executive report of this deadly storage tank fire caused by one of the workers was smoking inside the storage tank while sludge and residual crude oil was being removed from the tank (HSE, 1989).

At 1:20 p.m. (1320 hours), the outside contractor observed a fire ring inside the storage tank surrounding all three workers. Two of the workers managed to escape; however, one was killed when he was severely burned and overcome by smoke. The fire escalated rapidly, with flames and smoke coming from the open manways.

The HSE concluded that the cause of the fire was smoking by one of the workers inside the storage tank. Apparently, it was common practice for the workers to remove their breathing apparatus and smoke when the supervisor was not looking or occupied. In this instance, one of the workers who had been smoking inside the tank dropped a lit cigarette onto the floor of the tank, where it ignited the residual crude oil inside the tank.

There is a link in the "Additional References" section of this chapter to a video documentary of the fire brigade and police searching the tank internals to recover the victim. It is interesting due to the challenges they faced, including the sludge remaining and continuing ignition due to pyrophoric material in the tank.

The HSE made several recommendations to prevent a recurrence, including:

- The workers disregarded basic safety procedures, resulting in a lit cigarette being dropped inside the storage tank. These rules must be communicated and enforced.
- Terminal rules required that all matches and lighters be surrendered at the security gate, but this was not enforced.
- The hazardous area of electrical classification, flameproofing, and identification and elimination of ignition sources should be improved.
- No forced air ventilation was used to eliminate flammable vapors. This should be included in work procedures.
- The contract staff was not informed of the toxic and flammable hazards while working inside the crude oil storage tank.

Additional Comments by the Author

- Site safety rules, like smoking in non-authorized areas, must be broadly communicated to all employees and contractors and must be enforced. Rules of this type should be conditions of employment, and violations should result in immediate termination.
- The no-smoking rule is another rule that should be considered and rigorously enforced as a "Life Saving Rule" (See Chapter 8 for additional information on "Life Saving Rules").
- The tank should have been gas tested for oxygen content (O_2), flammables (% LEL), and hydrogen sulfide (ppm) before entry and periodically while work was in progress. Better yet, in today's environment, we would expect to use the continuous gas testing process with real time monitoring.

  Expected results should target a minimum O_2 of 19.5%, a maximum LEL of 10%, and a maximum H_2S of 20 ppm (Referencing OSHA requirements, of course, these would be required to meet UK requirements).
- The potential for pyrophoric material should have been considered during the work planning process, and precautions should have been laid out before the work began, including periodic wetting of the tank interior to prevent material drying and the potential for ignition.

I realize that tank cleaning rules and standards have changed dramatically since this incident, primarily due to similar incidents in the industry. Due to these improvements, placing workers inside a storage tank while the tank still contains crude oil or other flammable hydrocarbons would be highly unusual. To help prevent an incident like this one, this work should be done from the tank manways using remotely operated equipment.

Additional References

British Health and Safety Executive (HSE), Fire in a Crude Oil Storage Tank, BP Oil Dalmeny, Scotland. June 11, 1989, https://www.hse.gov.uk/comah/sragtech/casebpdalment87.htm#:~:text=Accident%20summary,series%20of%20219%20support%20pillars.

Facebook, Video Documentary of Dalmeny Oil Terminal Tank Fire, https://www.facebook.com/museumoffireedin/videos/dalmeny-tank-farm-1161987/525882744459531/.

Gazetteer for Scotland, Dalmeny Tank Farm, https://www.scottish-places.info/features/featurefirst18053.html.

Series V. Explosion, Fire, or Loss of Life Due to Inadequate or No Gas Testing

Incident 29: Explosion, Fire, or Loss of Life Due to Inadequate Gas Test, or Failure to Gas Test

SABIC Innovative Plastics, LLC, Ottawa, Illinois

An accidental release of butadiene occurred inside a decanting tank, resulting in a fire.

One employee was seriously injured in the incident.

The incident occurred on September 5, 2023.

At this site, workers obtained a confined space entry permit to access a decanting tank and remove an internal baffle with hand tools. One worker entered the tank, while another stayed outside as a confined space attendant. The tank was monitored using a five-gas detector.

Shortly after starting removal with the battery-powered tool, flammable vapors ignited inside the vessel, causing serious burn injuries to the entrant. The entrant was able to escape the vessel but needed treatment at the local hospital. The fire was extinguished by two other SABIC employees.

This incident occurred during a maintenance shutdown in the latex production area. The SABIC procedure required that the vessel be cleaned before entry with high-pressure water to remove any residual material, including accumulated solids. This step was not performed. Residual solids remained in the vessel and possibly contained some butadiene, leading to the release of flammable vapors, which were ignited, possibly by the battery-operated tool or sparks.

Even though continuous gas monitoring was in place during vessel entry, the end of the air.

The monitor hose was not positioned near the bottom of the tank where the solids were, which likely prevented the detection of vapors by the five-gas detector. SABIC concluded that the work inside the vessel probably disturbed the residual solids, releasing the butadiene vapor. Refer to Figure 11.47, which shows the sludge remaining at the bottom of the tank.

Probable Cause

The following is a quote from the CSB concerning this incident, including "inadequate flammable gas monitoring":

> "Based on SABIC's investigation, the CSB determined that the probable cause of the incident was performing work in a confined space containing a flammable atmosphere. Inadequate flammable gas monitoring, the use of a

US Chemical Safety and Hazard Investigation Board (CSB). Public Domain.

Figure 11.47 Interior of SABIC Tank Containing Solids.

> battery-powered tool, and employees' being unaware that SABIC had a procedure requiring the tank to be cleaned before entry contributed to the incident."

Author's Comments

This example illustrates how a failure in procedure execution directly caused an incident that resulted in severe burn injuries. There was a procedure in place that, if properly followed, should have prevented the presence of flammable vapors in the vessel. This procedure mandated cleaning the vessel with high-pressure water to remove any residual product, including solids, before issuing an entry permit.

However, inadequate gas testing procedures and failure to follow the procedures also contributed to this event. As outlined in Chapter 6, "Key Elements for Effective Gas Testing," it is the gas tester's responsibility to ensure that all areas within the monitored zone and/or vessel are tested to confirm that no flammable gases or toxics are present.

Additional Reference

US Chemical Safety and Hazard Investigation Board (CSB), Incident Reports/Events Reported to the CSB Under the Accidental Release Reporting Rule, Volume 2 March 12, 2025 (Page 39), SABIC Innovative Plastics, LLC, Ottawa, Illinois, https://www.csb.gov/assets/1/6/incident_reports_volume_2_2025-03-12_r1.pdf.

Incident 30: Explosion, Fire, or Loss of Life Due to Inadequate Gas Test, or Failure to Gas Test

International-Matex Tank Teminals, St. Rose, Louisiana

Flammable vapor within a storage tank ignited, resulting in an explosion and a major fire at this facility in St. Rose, Louisiana.

This incident occurred on April 3, 2023. It resulted in serious injuries to two contract employees and over $15 million in property damage.

On August 29, 2021, a Category 4 hurricane with winds of 150 mph hit South Louisiana. Shortly afterward, IMTT discovered that the storm had blown off an access hatch on a large storage tank, which provides access to the tank's internal floating roof.

In preparation for hot work on the storage tank, IMTT reviewed the inventory records, which showed the tank was empty. However, the tank's piping was not locked out, and no manways were removed to verify that the tank was empty and free of residual flammable liquids.

IMTT issued a hot work permit for contractors to grind and weld on the tank to install a new hatch on the storage tank. The tank was gas-tested around the access ladder and up to two feet inside the tank's opening, and no flammable vapors were found in this area. IMTT also required workers to cover the tank opening with a fire blanket and mandated continuous air monitoring on the tank's roof while the work was ongoing.

While the two contract employees were working on installing the new hatch near the top of the tank, it exploded, causing a large fire in the tank. The two employees were airlifted to the hospital for treatment of their serious burns. Other emergency responders extinguished the storage tank fire. Figure 11.48 is from the U.S. Chemical Safety and Hazard Investigation Board (CSB) illustrating the damage caused by the tank, believed to be due to ignition of hydrocarbons remaining in the tank while hot work was being performed.

Probable Cause

The following is a quote from the CSB concerning this incident, including "ineffective air monitoring practices to identify the presence of flammable hydrocarbon vapor contributed to the incident."

US Chemical Safety and Hazard Investigation Board (CSB). Public Domain.

Figure 11.48 Destroyed storage tank at the IMTT Terminal, St. Rose, Louisiana.

> "Based on IMTT's investigation, the CSB determined that the probable cause of the incident was the presence of flammable hydrocarbon vapors inside the tank while hot work (grinding or welding) was being performed on the tank's roof. Not thoroughly draining, cleaning, and purging the tanks to remove the flammable material before starting the hot work contributed to the incident. Additionally, ineffective air monitoring practices to identify the presence of flammable hydrocarbon vapor contributed to the incident."

Author's Comments

This is another instance where failure to properly prepare the vessel for hot work, including verifying the vessel's isolation from other piping and/or vessels in service, and confirming the vessel's contents, can lead to issues. As outlined in Chapters 3, 4, and 7, and in other discussions throughout this book, the vessel where entry or hot work is being performed should be isolated from other equipment, preferably air-gapped at all flanges and connections, and properly purged to eliminate flammable or combustible vapors.

In this incident, procedure failures combined with ineffective gas testing led to an explosion, fire, and serious, painful injuries to personnel. When conducting gas testing, the expectation should be to test all areas inside the vessel and the surrounding area to confirm that the vessel is free of hydrocarbons and toxic vapors. In the case of storage tanks with floating roofs, this means testing at all manways, including the floating roof pontoons.

Additional References

NOLA News/April 3, 2023, Fire at IMTT Fuel Terminal in St. Rose Sends Two to Hospital, https://www.nola.com/news/environment/fire-at-imtt-fuel-terminal-in-st-rose-sends-two-to-hospital/article_a7688ce2-d25a-11ed-9204-57dcc6fafc19.html.

US Chemical Safety and Hazard Investigation Board (CSB), Incident Reports/Events Reported to the CSB Under the Accidental Release Reporting Rule, Volume 2 March 12, 2025 (Page 39), IMTT Tank Terminals, St. Rose, Louisiana, https://www.csb.gov/assets/1/6/incident_reports_volume_2_2025-03-12.pdf.

Incident 31: Explosion, Fire, or Loss of Life Due to Inadequate Gas Test, or Failure to Gas Test

Dow Chemical Company

Hahnville, Louisiana facility (Operated by Union Carbide Company)

A flash fire occurred in a product purge vessel while workers were using a grinder to remove bolts from the vessel's top head flange.

Four workers were seriously injured, requiring hospitalization.

Incident date: February 3, 2023.

This incident was documented in Volume 2 of the CSB incident summary. It is included here to highlight the importance of gas testing in preventing accidents. The incident involved a flash fire and four serious. Dow reported it to the CSB, and the information provided by Dow is included in this report. A link to the CSB report is available in the "Additional References" section below.

The polyethylene unit at this facility was shut down on January 29 for planned maintenance, and operators isolated the product purge vessel and purged it in preparation for workers to replace the internal filter elements. No hot work was planned for this vessel until the workers found that several of the bolts for the top head were corroded, and removal using conventional tools was not working. A hot work permit was requested and received from Dow to allow the cutting of the remaining bolts using a grinder.

On February 3, when workers were removing the few remaining bolts using the grinder, the hot metal and sparks ejected from the grinder ignited a flammable mixture of hydrocarbon vapor and air inside the product purge vessel. As a result, four of the workers who were near the top head received serious burn injuries and were hospitalized for treatment.

Dow investigated this incident and found that the purge vessel was connected to the flare system, which allowed the safe venting of unreacted gases when the unit was in service. When operators isolated the vessel, a single isolation valve was closed, and since no hot work was planned, no isolation blinds or air gaps were provided for isolation. Other valves were available for isolation; however, only the single valve was closed. The investigation revealed that this single valve leaked hydrocarbons into the product purge vessel, providing the fuel source for the flash fire. Figure 11.49 is from the investigation report completed by the U.S. Chemical Safety and Hazard Investigation Board following the flash fire at Dow Chemical's Hanville, LA. facility, which was operated by Union Carbide Company.

The Dow investigation reported that the facility has an equipment isolation procedure that requires isolation blinds and air gaps to ensure no connections remain to live equipment. The procedure also requires the use of nitrogen to purge any remaining hydrocarbon vapors from the vessels. However, not all elements of this procedure were followed in this case. As noted above, only one valve was used for isolation; no blinds or air gaps were provided. Noteworthy for this chapter is that a gas test was conducted for the atmosphere around the vessel, which found the area to be gas-free. No gas testing was conducted inside the vessel, which, if one had been done, certainly would have indicated the presence of a flammable mixture.

US Chemical Safety and Hazard Investigation Board (CSB). Public Domain.

Figure 11.49 The Dow product purge vessel after the flash fire.

Probable Cause

Note that the following paragraph is a quote from the CSB report:

> "Based on Dow's investigation, the CSB determined that the probable cause of the flash fire was performing hot work (grinding) to cut flange bolts on a pressure vessel containing a flammable atmosphere. The ineffective application of the hot work policy contributed to the incident by relying on a single isolation valve to prevent flammables from entering the vessel from the flare system and not performing combustible gas testing of the flammable atmosphere within the vessel before permitting this work. Had combustible gas testing of the atmosphere within the vessel been conducted before permitting the work, this incident likely could have been prevented."

Author's Comments

This incident and its causal factors are straightforward. Procedures existed for isolating process vessels for opening and hot work, but they were not followed.

The CSB report focused not only on vessel isolation, specifically on ensuring the provision for creating air gaps, but also on vessel purging of any remaining hydrocarbons or chemicals, and on gas testing inside the vessel before allowing hot work, whether flame or, as in this case, non-flame hot work.

In this incident, the area around the outside of the vessel was gas tested; however, the interior of the vessel was not tested. A gas test of the interior would certainly have revealed that vessel isolation was not adequate and that hydrocarbons were present in the vessel. Based on a valid gas test of the vessel interior, a hot work permit should not be issued.

Additional Reference

US Chemical Safety and Hazard Investigation Board (CSB), Incident Reports/Events Reported to the CSB Under the Accidental Release Reporting Rule, Released: March 12, 2025 (Page 27), https://www.csb.gov/assets/1/6/incident_reports_volume_2_2025-03-12.pdf.

Incident 32: Explosion, Fire, or Loss of Life Due to Inadequate Gas Test, or Failure to Gas Test

Westlake Corporation, LLC/Westlake, Louisiana

January 26, 2022

A tank explosion occurred when workers were dismantling a nitrogen pressure regulator for repairs. A one-million-gallon ethylene dichloride storage tank was destroyed.

No injuries occurred, but property damage was extensive.

This incident was documented in Volume 3 of the CSB incident summary. It is included here to highlight the importance of gas testing in preventing accidents. The incident involved a catastrophic explosion in a one-million-gallon (~24,000 bbl) ethylene dichloride storage tank. Westlake, Louisiana, reported it to the CSB, and the information provided by Westlake is included in this report. A link to the CSB report is available in the "Additional References" section below.

On January 26, 2022, workers were removing corroded piping related to a nitrogen pressure regulator. The regulator was designed to supply a nitrogen blanket on the ethylene dichloride cone roof storage tank to prevent the formation of a flammable mixture. Workers were in the process of removing the pressure regulator and associated piping for repairs when the incident occurred.

Tank inspection had previously identified multiple holes in the tank roof caused by corrosion and severe corrosion on the nitrogen piping connected to the pressure regulator. The day before the explosion, the tank was taken out of service to repair the corroded holes in the roof. Most of the ethylene dichloride was removed from the tank, although some residual products remained, and flammable vapors continued due to the remaining ethylene dichloride in the tank.

Westlake had operators issue a work permit to allow the workers to remove the nitrogen pressure regulator for repairs. The workers were in the process of removing the piping using a battery-powered reciprocating saw to remove the bolts on the piping flanges when one of the workers noticed a small flame inside the tank that was visible through one of the holes in the tank roof. As soon as the fire was noticed, all the workers quickly evacuated from their worksite on the tank roof to a nearby maintenance shop and notified the control room operators about the fire. At approximately 10:45 a.m., approximately three minutes after the workers reported the fire, the tank exploded.

When the tank exploded, the shell completely separated from the tank floor, resulting in the total loss of approximately 3,000 pounds of ethylene dichloride remaining in the tank, which burned in the resulting fire. See Figure 11.50 for images of the Westlake storage tank fire. Figure 11.51 illustrates the significant damage resulting from the tank explosion.

Following this major safety incident, Westlake investigated the cause of the explosion and fire. They found that a flammable environment existed inside the storage tank due to piping misalignment, which allowed ethylene and oxygen to flow into the tank from a separate but connected process unit for about two weeks before the tank was isolated to repair the roof (to patch holes in the roof). After three minutes of burning on the roof's surface, the fire was drawn into the tank, causing the explosion. Westlake also reported that the work permit issued to the workers did not authorize the use of a battery-powered reciprocating saw or other tools that could create a spark.

US Chemical Safety and Hazard Investigation Board (CSB). Public Domain.

(a) KATC 3

US Chemical Safety and Hazard Investigation Board (CSB). Public Domain.

(b) KPLC 7

Figure 11.50 Smoke from the January 26, 2022, explosion at Westlake Corporation in Westlake, Louisiana.

Probable Cause

Note that the following paragraph is a quote from the CSB report:

> "Based on Westlake's investigation, the CSB determined that the probable cause of the incident was a flammable atmosphere created inside the tank from a piping alignment that allowed ethylene and oxygen vapor to flow from an interconnected process unit into the tank. The flammable vapor

US Chemical Safety and Hazard Investigation Board (CSB). Public Domain.

Figure 11.51 The tank after the explosion.

was ignited from a spark created by using a battery-powered reciprocating saw to cut corroded bolts during a maintenance activity."

Author's Comments: The Value of Gas Testing

The CSB report states that workers were issued a work permit to patch holes in the tank roof and repair the corroded piping related to the nitrogen pressure regulator. However, the report does not mention a gas test or whether one was considered. Westlake reports that the work permit did not authorize the use of any arcing or sparking tools or equipment. This is an important point, and the work permit is exactly where the authorization to use or not use electrically powered equipment or any other arcing or sparking devices should be specified. However, this should also be discussed during the joint job-site visit, including the identified hazards and how they will be controlled as the work progresses. The joint site visit is where the work permit should be issued, in the field and at the location where the work will be performed.

The CSB report did not specify what type of work permit was issued for repairing the corroded tank roof or removing the nitrogen tank blanketing piping. Westlake stated they did not authorize the use of a battery-powered reciprocating saw. Most petroleum refineries and petrochemical plants have separate permits for cold work and hot work. Some facilities have at least two hot work permits

– one for flame-work, such as burning or welding, and another for non-flame work such as grinding. The key point is that any hot work should require a hot work permit, which includes a gas test for flammables before starting any activity that can ignite a flammable mixture.

The flash point of ethylene dichloride is 55 °F (13 °C), making it classified as flammable. Above this temperature, any remaining product in the storage tank would vaporize, creating a flammable or explosive atmosphere inside. The CSB report states that the tank contained residual ethylene dichloride. See the "Additional References" section for a link to the ethylene dichloride SDS.

In this situation, performing a gas test would have provided vital data to evaluate the hazards involved. It would have confirmed the presence of flammable gas and indicated that precautions to prevent ignition were necessary. This should lead to additional safety measures beyond simply banning the use of arcing devices. For example, many explosions result from static electricity, which can be present in the fluids or introduced through tools or conveyances used during the transfer of liquids or gases. Ethylene dichloride is considered a nonconductor, making it susceptible to static discharge. These hazards are typically controlled by equipment or transfer mediums through electrical bonding and grounding to prevent static electricity from igniting a spark. My point is that all potential sources of ignition should be considered, and precautions developed to prevent ignition of flammable vapors.

This was truly a close call, and I think about the scenarios that could have easily turned out to be catastrophic. What might have happened if the worker had been distracted and hadn't noticed the small fire through the corroded hole in the tank roof, or if the workers had failed to recognize the potential consequences and stayed on the tank's roof? What could have happened if the explosion had occurred while the response team was near the bottom of the tank, preparing to extinguish the fire? Certainly, the damage was significant, but I believe you would agree it could have been far worse. This should serve as a crucial lesson for the entire industry. We should never dismiss this as just a tank explosion or fire – it could have easily resulted in a major loss of life.

Additional References

US Chemical Safety and Hazard Investigation Board (CSB), Incident Reports/Events Reported to the CSB Under the Accidental Release Reporting Rule, Released: July 22, 2025 (Page 25), https://www.csb.gov/assets/1/6/incident_reports_volume_3_2025-07-22.pdf.

Westlake, Ethylene Dichloride Safety Data Sheet, https://www.westlake.com/safety_guidelines/Geismar_Safety_Guidelines/Ethylene%20Dichloride%20(EDC).pdf.

Incident 33: Explosion, Fire, or Loss of Life Due to Inadequate Gas Test, or Failure to Gas Test

TE Products Pipeline Company, LLC (TEPPCO)
McRae Product Terminal, Garner, Arkansas
May 12, 2009; three fatalities

This tragic accident involved the explosion of an out-of-service gasoline storage tank at the McRae Product Terminal. The explosion occurred when contract personnel were cutting the internal floating roof with a flame cutter. An independent investigation revealed that gas testing for this project was severely inadequate.

Tank 1303 was out of service and prepared for the installation of a new gauge pole in the external floating roof storage tank. Documentation for this project highlights three key steps in the execution: (i) TEPPCO was to isolate the storage tank from other operating or in-service equipment. (ii) The second step was to clean the tank and make it safe for hot work. TEPPCO hired an outside contractor to clean the tank floor; however, the pontoons were not cleaned. (iii) The third step involved another outside contractor entering the tank to install the new gauging pole, a task known to require internal hot work such as burning and welding.

The TEPPCO procedures stated that the storage tank must be free of flammable materials before issuing a work permit. A job plan for this project was created, specifically listing gas testing requirements for all tank internals, including the floating roof pontoons. The plan called for an initial gas test and continuous gas monitoring to prevent the buildup of flammable vapors during the hottest part of the day or from leaks in the overhead pontoons. This form was signed by TEPPCO personnel and the contractors involved in the project. However, only an initial gas test of the tank internals was performed, recorded at 7:00 a.m., though the location and number of readings are unclear. No additional testing was conducted before the explosion at 2:17 p.m.

The independent investigation found that hazardous vapors ignited when a welder cut the internal floating roof with a flame cutter. It was also noted that the gas testing monitoring equipment was easily accessible, but it was not used again after the initial test at 7:00 a.m. There was no subsequent test and no continuous monitoring before the explosion. The report further confirmed that no gas testing monitoring equipment was found in the debris, indicating that continuous gas testing was not performed as required by the project plan and TEPPCO standards. Figure 11.52 is from the U.S. Chemical Safety and Hazard Investigation Board investigation report of this tragic incident which was believed to be caused by inadequate gas testing before conducting hot work.

Author Comments

In an incident like this one, I rely on the OSHA standards. OSHA 29 CFR 1910.146 (Permit-required confined spaces) clearly details the requirements for entering a confined space like the one in this incident.

US Chemical Safety and Hazard Investigation Board (CSB). Public Domain.

Figure 11.52 Image of the large gasoline storage tank at the TEPPCO Pipeline Terminal that was destroyed in this tragic explosion.

1910.146(c)(5)(ii)(C) clearly states:

> "Before an employee enters the space, the internal atmosphere shall be tested, with a calibrated direct-reading instrument, for oxygen content, for flammable gases and vapors, and for potential toxic air contaminants, in that order. Any employee who enters the space, or that employee's authorized representative, shall be provided an opportunity to observe the pre-entry testing required by this paragraph."

1910.146(c)(5)(ii)(F) also clearly states the requirement for retesting:

> "The atmosphere within the space shall be periodically tested as necessary to ensure that the continuous forced air ventilation is preventing the accumulation of a hazardous atmosphere. Any employee who enters the space, or that employee's authorized representative, shall be provided with an opportunity to observe the periodic testing required by this paragraph."

The OSHA Confined Space Standard also requires other precautions and mitigation measures to be implemented during entry. These are not mentioned in the report, so whether they were implemented on this project is unclear. These include the following (not intended to be a complete list):

- Gas test requirements
- Entry permit and requirements
- Forced air ventilation
- Barriers and shields
- Rescue plan and equipment
- Entry attendant (and attendant requirements and duties)
- Training (duties of entrants and duties of attendants, and duties of entrant supervisors)

Having said this, I am a strong advocate for continuous gas monitoring. We must acknowledge that conditions change during the execution of a job or project in the field. These changes include higher ambient temperatures caused by daytime heating. Daytime heating vaporizes any residual hydrocarbons that may be present and can create flammable mixtures. Additionally, leaks or releases may occur at neighboring units or operations, which can negatively impact entrants. Throughout my career, I have encountered many of these issues, and continuous gas monitoring is performed in real-time, providing audible and visual alarms to warn workers of any changes in conditions.

For example, I worked as an independent consultant on a major construction project at a West Coast Refinery. I authored the project safety plan, which included continuous gas monitoring, and had the opportunity to observe its implementation. This project involved installing large new equipment very close to other operating process equipment. A massive excavation was necessary for the foundation and other underground components. We contracted for four continuous gas monitors, one at each corner of the excavation, that tracked oxygen, LEL, H_2S, and HF, each equipped with audible and visual alarms. The project was completed without incidents and with virtually no disruption to the original schedule.

As I reflect on the TEPPCO tank explosion, I can't help but think that if they had used something like this, the incident would not have occurred. Of course, the pontoons constitute separate confined spaces, and as such should have been gas tested in both the initial and follow-up tests to ensure that flammable vapors are not being generated from trapped hydrocarbons.

CSB: Storage Tank Safety Bulletin

The CSB published a Safety Bulletin in February 2010 titled "Seven Key Lessons to Prevent Worker Deaths During Hot Work in and Around Tanks."

The following paragraph is taken directly from this bulletin.

> "The CSB has identified over 60 fatalities since 1990 due to explosions and fires from hot work activities on tanks. Hot work is defined as 'work involving burning, welding, or a similar operation that is capable of

initiating fires or explosions.' Hot work also includes other activities with the potential to create a source of ignition, such as cutting, brazing, grinding, and soldering. Workers are potentially at risk, not only in the oil and gas industry, where flammables are handled regularly, but also in many other sectors within general industry, such as food production, paper, and wastewater treatment."

"This bulletin, however, focuses on seven key lessons drawn from incidents to emphasize recurring safety issues that deserve special attention during hot work operations. For a broader discussion of general work practices for hot work, the reader should consult these standards, as well as other informative sources listed in the references section of the bulletin. This bulletin summarizes 11 accidents – nine of which the agency has investigated – to highlight seven key lessons that were found to be applicable to all or most of the incidents, especially the need for effective hazard assessment and proper monitoring of potentially flammable air concentrations in work areas."

Additional References

EHS Today, CSB Issues Safety Bulletin on Hazards of Welding, Hot Work, Laura Walter – March 15, 2010, https://www.ehstoday.com/safety/article/21908874/csb-issues-safety-bulletin-on-hazards-of-welding-hot-work.

US Chemical Safety and Hazard Investigation Board (CSB), Seven Key Lessons to Prevent Worker Deaths During Hot Work in and Around Tanks, https://www.csb.gov/assets/1/17/csb_hot_work_safety_bulletin_embargoed_until_10_a_m__3_4_101.pdf?14329.

Incident 34: Explosion, Fire, or Loss of Life Due to Inadequate Gas Test, or Failure to Gas Test

Pennzoil Refinery, Rouseville, Pennsylvania

A storage tank exploded, and a rapidly spreading fire occurred in adjacent tanks and other refinery areas.

October 16, 1995; five fatalities, two others injured, one critically. This incident also led to a public evacuation and significant plant damage.

The EPA investigated this incident and published a detailed report. Included in the report are information and excerpts from the EPA report. This report is abbreviated to fit within this format. Refer to the full report if additional information is desired. There is a link in the "Additional References" section of this report.

Although it happened a long time ago, I've included this incident here because it was a very tragic event, and there are many important lessons to learn and share from what occurred.

This catastrophic explosion and fire occurred around 10:15 a.m. on October 16, 1995, when welders were working on a stairway between two waste liquid storage tanks (Tanks 487 and 488). These tanks contained a mixture of liquid waste, hydrocarbons, and water. Before beginning work, a proper gas test was conducted, as required by site procedures and OSHA.

During the initial explosion, both storage tanks "rocketed" off their foundations, releasing burning product over a wide area. The EPA reported that the flammable vapors in Tank 487 ignited, causing a rapid increase in pressure inside the tank. This pressure led the tank to fail at the bottom seam and be hurled into the air. The burning liquid from Tank 487 ignited vapors in the neighboring tank (Tank 488), which also failed at the bottom seam and was similarly "rocketed" from its foundation, releasing burning liquid and spreading the fire further. Because there were no containment berms around the tanks, the burning liquid from both tanks quickly spread throughout the refinery, creating a very large fire.

The burning liquid spread across construction zones, storage trailers, a contractor break trailer, and various other product tanks. The flames from the released liquids ignited several tanks containing naphtha and fuel oil. Additionally, loud explosions occurred due to the rupture of compressed gas cylinders and other sealed containers. Some debris was thrown beyond the refinery's fence line. The spreading fire prompted the evacuation of the town of Rouseville, local schools, and the refinery itself.

Unfortunately, the explosion and fire caused the deaths of five workers, including the welders, and injured one other person. The refinery was badly damaged by the explosions and fires.

Facility and Background Information (As Determined by the EPA Investigation)

The first explosion happened at Tanks 487 and 488, two tanks used for storing waste mixtures of water, oil, or other hydrocarbons. These tanks were designed to hold wastewater and slowly drain it into the wastewater treatment system to prevent overloading the process. The waste mixtures were usually offloaded into the storage tanks from vacuum trucks that collected them from various parts of the facility.

These two similar tanks measured 30 feet in diameter, with 25-foot-high sidewalls and five-foot domed roofs. Positioned about four feet apart, they shared a common stairway to ease tank gauging and sampling. The tanks did not have individual dikes but were enclosed with other storage tanks to prevent spills from reaching the nearby creek.

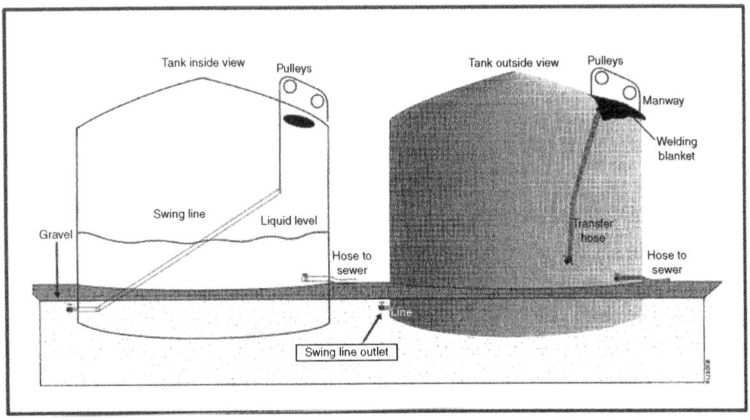

Figure 11.53 Image of Tanks 487 and 488, at Pennzoil Products Company, Rouseville, Pennsylvania. Note that the internal "swing line" and the swing line outlet, operated by pulleys and ropes. Also, observe the open roof manway, covered by a welding blanket, with the vacuum truck hose extended to grade. US Environmental Protection Agency (EPA). Public Domain.

The waste mixtures were pumped into the tanks from vacuum trucks that collected waste liquids from drains, equipment, or other locations around the site. The tanks were designed as settling tanks to separate the collected waste into natural phases of hydrocarbons and water, allowing the water to be drained and the hydrocarbons skimmed from the tank. This process enables the hydrocarbons to be processed and reclaimed. There were no connections between these two tanks and other storage tanks at the facility.

The investigators found that, when the explosion, a 3-inch-diameter transfer hose was draped over the side of Tank 487 and into an open access area on top of the tank. The other end of the hose was at ground level, where vacuum trucks could unload collected oil and waste into the tank. The hose was secured beneath the stairway with wire. The tank was equipped with a "swingline" inside, operated by a cable and pulley system that allowed the pump suction to be raised and lowered, enabling the removal of hydrocarbons from the top layer or water from the bottom. At the time of the explosion, the swingline on Tank 487 was located in the vapor space above the liquid level.

See Figure 11.53 for views of Tank 487, which shows the transfer hose, swingline, pulley, and drain valve.

Chemical Composition of the Product Contained in the Tanks at the Time of the Explosion

It is impossible to determine the exact chemicals and their concentrations in the tanks at the time of the explosion, as well as the exact amounts or levels in each tank. Due to the nature and use of the tanks, the hydrocarbon composition could vary widely depending on the slop oil being recovered. The investigators found that the last time the tanks were gauged, Tank 487 held about 10,000 gallons and Tank 488 held about 116,000 gallons. No other information was available on the hydrocarbon volume or water layers.

What is known is that several days before the explosion, a vacuum truck released an unknown amount of methyl ethyl ketone (MEK) into one or possibly both of the tanks. MEK is a well-known flammable liquid and is only partially soluble in water. It also has a very low flash point (19 °F or −7 °C) and a narrow flammable range, with 1.4% LEL to 44% UEL, making it highly flammable. Additionally, MEK's density is two and a half times heavier than air, causing vapors to quickly settle at lower levels or on the ground.

The Explosion and Fire

A Pennzoil safety officer issued a hot work permit around 7:40 for two Pennzoil employees to weld a handrail to the stairway of storage Tanks 487 and 488. Pennzoil safety technicians had already inspected and approved the setup for welding. The permit required that all manways on top of both tanks be covered, the welding machine be grounded near the welding area, and that a fire watch be present at all times at ground level. A fire watch is a person designated to watch for small fires that may occur from welding slag or spatter dripping from the work area; the fire watch must have fire extinguishing equipment readily available and be trained to use it. Additional precautions to prevent fires included noting that no welding should occur near the top of the tanks (where the manways are located). The permit also mandated measuring the area for combustible vapor.

The area around the stairway was prepared for hot work. A welding machine was set up near the work site, the stairway handrail was secured in place, and welding canvas blankets were laid down to collect sparks and help prevent fires in the area. Gas tests were conducted at ground level and showed no evidence of combustible vapor in the welding zone.

One welder started arc welding on the stairs using an internal combustion welding machine, while the second welder acted as a fire watch on the ground. The welder was told not to weld on the handrail closest to the roof of the tank because it was too near the tank openings.

Both welders took a break around 9:30 a.m. after welding the handrail in place. They returned to the worksite around 10:00 a.m. and tried to resume their work, but unfortunately, the welding machine wouldn't restart. Around 10:10 a.m., two employees in a maintenance truck helped jump-start the welding machine.

Almost immediately afterward, around 10:15, an explosion occurred in Tank 487, followed by a second explosion in Tank 488 less than a minute later. The investigators could not determine if the welders had resumed welding operations when the explosion happened. However, witnesses reported seeing one welder on the platform between the two flights of stairs and the other at the bottom of the stairs.

Each storage tank failed catastrophically at the bottom seam where the vertical sidewall was welded to the bottom floorplate. This failure caused each tank to lift, with the trajectory angling in opposite directions from their original positions. Tank 487 lifted off its base and landed about 20 feet to the west. Tank 488 landed approximately 50 feet to the east after passing over another storage tank and a pipe rack. Naturally, the burning contents of each tank were released, creating a wave of burning hydrocarbons that engulfed the entire area.

Refer to Figure 11.54, which illustrates the overpressure and tank trajectories of Tanks 487 and 488 after the explosion.

Causes of the Accident (Abbreviated Here)

The investigation team identified that the immediate cause of the fire was the ignition of flammable vapors in storage Tank 487. Although the team could not determine the exact mechanism, they determined at least two likely scenarios:

- Undetected flammable vapors emitted from Tank 487 were ignited by an ignition source, which then flashed back into the tank.
- An electrical discharge in Tank 487, generated by the arc welding, ignited flammable vapors in the tank.

The Investigation Team Identified the Following Four Potential Sources of Flammable Vapor

Vapor Escaping from Open Manway

Placing a hose to discharge liquids from vacuum trucks through the open manway into Tank 487 prevented the manway cover from fully sealing the opening. Although a welding blanket was draped over this opening, vapor could have escaped into the surrounding area because the blankets are not designed to provide a vapor-tight seal.

Vapor Escaping from Swingline

The swingline might have allowed vapor to escape. The swingline section inside the tank was located above the liquid level in the vapor space. Investigators found after the accident that the outside valve near the tank's base, which was connected to the swingline inside, was open.

Vapor Escaping from Transfer Hose

Vapor from the tanks could have escaped through the open, 3-inch transfer hose running from inside the storage tank, through the manway, to the ground outside the tank.

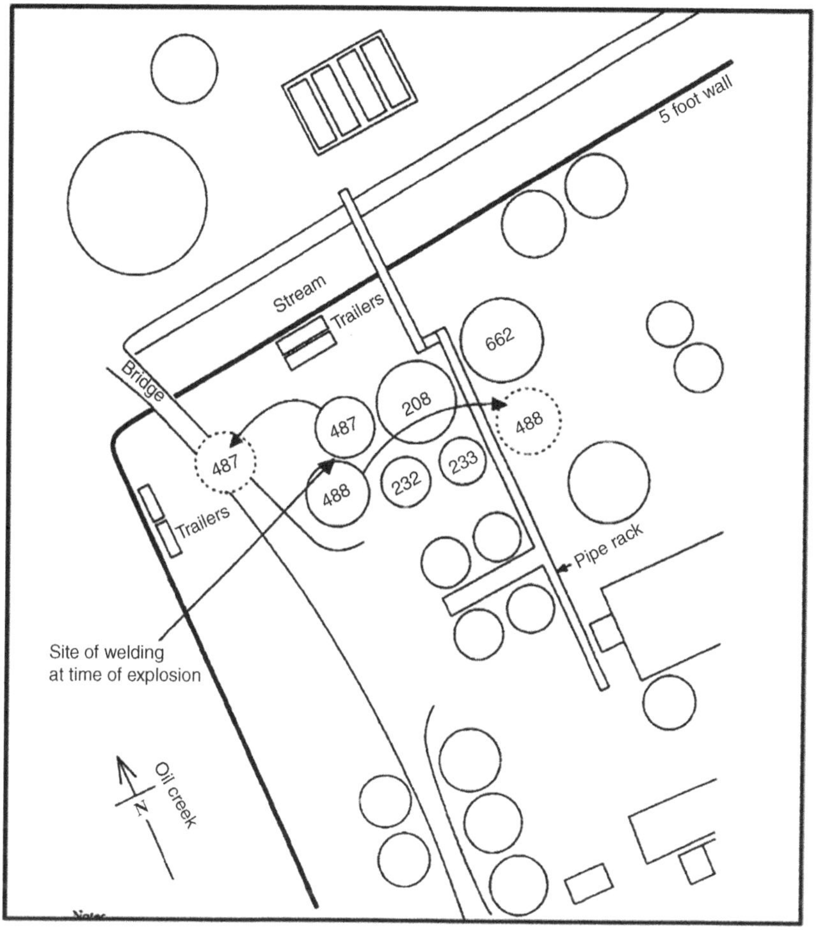

Figure 11.54 Illustration of Tanks 487 and 488 overpressure and trajectory after the explosion. US Environmental Protection Agency (EPA). Public Domain.

Vapor Escaping from Holes in Tank

After the explosion and fire, investigators found holes up to a quarter inch in diameter, likely caused by corrosion, near the top of the tank.

The Investigation Team Also Identified the "Likely" Ignition Sources

The investigation team determined that the most probable sources of ignition were the welding operation, the welding machine's internal combustion engine, and electrical arcing from the welding equipment or grounding cables.

US Environmental Protection Agency. Public Domain.

Figure 11.55 Pennzoil Products Company, Tank 487 and 488 After Explosion (October 16, 1995). Tank 487 is in the background.

Design of the Storage Tanks

- Both storage tanks failed at the bottom seam, creating a massive loss of containment and spreading fire to other parts of the facility.
- Cone roof design storage tanks are typically designed with a relatively weak top roof seam. With this design, if an internal explosion occurs inside the storage tank, the roof seam is designed to fail, releasing the pressure in the tank. This prevents the loss of containment that can occur when the bottom seam fails.

Figures 11.55 and 11.56 show the damaged and dislocated storage tanks after the devastating explosion and resulting fires.

Key Learnings

Follow Up Gas Testing or Continuous Gas Testing

The importance of ongoing or additional gas testing cannot be overstated. The EPA investigation report noted that the initial gas test was conducted before daytime heating began, and no subsequent testing was done, such as after the break before resuming work around 10:00 a.m.

It is notable that between 7:00 a.m. and 10:00 a.m., the air temperature at Pittsburgh International Airport (about 75 miles south of Rouseville) rose

US Environmental Protection Agency. Public Domain.

Figure 11.56 Pennzoil Products Company, Tank 488 After Explosion (October 16, 1995).

approximately 9 °F, from 39 °F to 48 °F. The ambient temperature in Rouseville likely increased by a similar amount. Additionally, direct sunlight on the storage tanks probably raised vapor temperatures within them, and the resulting thermal expansion may have pushed flammable vapors out of tank openings.

Continuous gas testing was not easily accessible when this incident happened, as it is now. Otherwise, I'm convinced the EPA team would have definitely pointed out its advantages.

I've said this before, and I'll say it again. Continuous gas testing can save lives. When used, we often don't realize how helpful it can be because no incidents occur to investigate, and no one is harmed.

Additional References

BW Technologies, RIG RAT III Wireless Multi Point Gas Detector with Sensors, https://www.proconsystems.com/wp-content/uploads/2018/01/RigRat-III_Datasheet.pdf.

Industrial Scientific, Understanding Real-Time Area Gas Monitoring: Importance and Benefits, https://www.indsci.com/en/blog/understanding-real-time-area-gas-monitoring-importance-benefits.

US Chemical Safety and Chemical Hazard Investigation Board (CSB), Seven Key Lessons to Prevent Worker Deaths During Hot Work in and Around Tanks; Effective Hazard Assessment and Use of Combustible Gas Monitoring Will Save Lives, https://www.csb.gov/assets/1/17/csb_hot_work_safety_bulletin_embargoed_until_10_a_m__3_4_101.pdf?14329.

US Environmental Protection Agency (EPA), EPA Chemical Accident Investigation Report, Pennzoil Product Company, Rouseville, Pennsylvania, https://nepis.epa.gov/Exe/ZyPDF.cgi/100039US.PDF?Dockey=100039US.PDF.

Appendix A

End-of-Book Quiz

1 What are some of the most common hazards of opening process equipment (OPE)?
Please list as many as you can.

2 Normalization of deviance is a term used to describe a process where a person, a work crew, or an entire organization has developed a process that deviates from well-established __________ or __________, and the deviance has occurred so many times that it has become the __________ __________ or the new norm.

3 How do we help prevent loss of containment when equipment is removed for several hours or days?

4 It is very important to ensure that all personnel understand that in an emergency or in the event any hazardous product is detected, all work should be __________ immediately and __________ only after the site is reevaluated and all clear is given. The work permit should be __________ before work restarts.

5 When preparing multiple heat exchanger trains for opening, what is the primary hazard that exists when opening flanges?

6 Please describe the process of hot bolting, for example, hot bolting a bank of heat exchangers.

7 How should flanges and access plates be identified before the issuance of a work permit to open, for example, for placing blinds in a long pipeline?

8 Which OSHA Standard requires employers to develop, implement, and follow detailed work procedures and processes for handling or processing hazardous chemicals?

9 What is the most hazardous time we are most vulnerable to a process safety incident? Please list examples.

10 Before opening equipment to the atmosphere, it must be ________, ________, ________, or chemically cleaned to remove the hydrocarbons and chemicals before air can enter it for the first time, possibly in several years.

11 When preparing for equipment startup, essentially, all of the ________ must be removed from the equipment before ________ are admitted into it, or the hydrocarbons can mix with the air, potentially creating an explosive mixture.

12 The OSHA requirements for entering a confined space, such as a process vessel, a rail car, or a tank truck, are readily available in the Standard ________.

Please select the most correct answer (s).

- **a** OSHA Standard 29 CFR 1910.147
- **b** OSHA Standard 29 CFR 1910.119
- **c** OSHA Standard 29 CFR 1910.95
- **d** OSHA Standard 29 CFR 1910.146

13 What is meant by the term life saving rules?

14 What is the expected consequence of violating a life saving rule?

15 Which of the following work rule violations should be considered a life saving rule?

Please select the correct answers.

- **a** Failure to wear a seat belt while driving inside the facility.
- **b** Failure to wear fall protection when working on an incomplete scaffold.
- **c** Failure to arrive at work at the prescribed time.
- **d** Failure to properly utilize the work permit system when opening a flange on a pipeline.
- **e** None of the above.
- **f** All of the above.

16 Please describe in your own words what the term normalized deviation means.

17 Following an actual event or even a near loss, the most important thing is ensuring that the generated __________ are documented fully and __________.

18 What are the common issues with preparing exchangers for blinding or opening when configured in parallel trains?
Please select the most correct answer.

- **a** It is difficult to verify that the product was completely drained from all exchangers.
- **b** Chemical cleaning may be ineffective due to fouling in some exchangers.
- **c** The exchangers may be physically located in a heavily congested work area.
- **d** One bank of the exchangers may be heavily fouled, and flushing may be ineffective, leaving hydrocarbons trapped in the exchangers.
- **e** None of the above.
- **f** All of the above.

19 First and foremost, we must recognize that we are vulnerable to the release of fluids or gases during all phases of the OPE process.

In our business, these fluids may be highly ____________ like petroleum naphtha or gasoline, they may be extremely volatile like ____________ ____________ ________ or may be highly toxic like __________ ____________.

20 The Pemex refinery in Deer Park, Texas, experienced what appears to have been an OPE incident in October 2024 when contract personnel were in the process of ____. This was a tragic incident where 2 contractors were killed, 13 more were injured, and a major release of hydrogen sulfide was released into the refinery and the surrounding communities.
Please select the most correct answer.

- **a** Removing access plates from a process vessel.
- **b** Removing a process pump to send to the shop.
- **c** Opening a process exchanger for maintenance.
- **d** Installing isolation blinds in a pipeline system.
- **e** None of the above.
- **f** All of the above.

21 What is a common hazard when opening a distillation column or a scrubber that has been in high hydrogen sulfide service?

22 Why is pyrophoric material a concern? What can happen due to pyrophoric materials in a distillation column?

23 What are the different types of isolation blinds, and how are they used when isolating process equipment?

24 Cone roof design storage tanks are typically designed with a relatively __________ top roof __________.

With this design, if an internal explosion occurs inside the storage tank, the roof seam is designed to __________, releasing the pressure in the tank. This prevents the loss of containment that can occur when the __________ seam fails.

25 Which of the following gas tests should be completed before entry into a permit-confined space?
Please select all that apply.

a Oxygen content (%).
b Hydrogen sulfide content (ppm).
c Flammables (LEL).
d Other toxic materials that may be associated with this process unit or area.
e All of the above.
f None of the above.

26 What are the acceptable limits set by the US Occupational Safety and Health Administration for the following?

Oxygen: Your Answer: ______________________________

Flammables: Your Answer: ___________________________

Hydrogen sulfide: Your Answer: ______________________

Appendix B

End-of-Book and End-of-Chapter Quizzes (with Answers)

End-of-Book Quiz (with Answers)

1 What are some of the most common hazards of opening process equipment (OPE)?
Please list as many as you can.

Answer(s):
The following is not intended to be a complete list.

The unexpected release of pressurized fluids or gases, including toxic gases, due to one or more of the following:

- The failure to properly isolate, drain, and depressurize the system or vessel.
- Leaking isolation valves.
- The fouling or plugging of process vents or drains.
- Workers working on the wrong piping circuit and opening a flange that is not prepared for opening or removal.
- Trapped liquids released during flange opening, allowing air to enter the system.
- High melting point materials form an internal product plug in the process system, which can melt when the line temperature rises above the product melting point, causing a release of the product.

2 Normalization of deviance is a term used to describe a process where a person, a work crew, or an entire organization has developed a process that deviates from well-established _________ or _________, and the deviance has occurred so many times that it has become the _________ _________ or the new norm.

Answer(s):
procedures; norms; accepted process

3 How do we help prevent loss of containment when equipment is removed for several hours or days?

Answer(s):
To prevent unexpected releases, piping or vessel openings that are expected to continue over several hours or days should be secured by installing slip blinds or blind flanges.

4 It is very important to ensure that all personnel understand that in an emergency or in the event any hazardous product is detected, all work should be ________ immediately and ________ only after the site is reevaluated and all clear is given. The work permit should be ________ before work restarts.

Answer(s):
stopped; restarted; reissued

5 When preparing multiple heat exchanger trains for opening, what is the primary hazard that exists when opening flanges?

Answer(s):
Even after significant efforts to clear the multiple trains, hydrocarbon liquids may still be trapped in the exchangers and can be released into the atmosphere or ignited when opening the vessels.

6 Please describe the process of hot bolting, for example, hot bolting a bank of heat exchangers.

Answer(s):
With the issuance of a hot work permit from a qualified individual, we can proceed with burning one bolt at a time, reinstalling the bolt and tightening it before moving to the next one. After all bolts are replaced, the hot work is discontinued, and the opening can proceed using the normal method without hot work.

7 How should flanges and access plates be identified before the issuance of a work permit to open, for example, for placing blinds in a long pipeline?

Answer(s):
After operations have verified isolation and confirmed zero energy, each flange or access plate that will be opened should be identified. This identification method should be used only for this purpose, and every flange must be labeled.

Each flange scheduled for opening should be reviewed during the start-of-job meeting between the permit issuer and the workers, and the workers should be informed that only marked flanges are to be opened.

8 Which OSHA Standard requires employers to develop, implement, and follow detailed work procedures and processes for handling or processing hazardous chemicals?

Answer(s):
The Process Safety Management Standard or 29 CFR 1910.119

9 What is the most hazardous time we are most vulnerable to a process safety incident? Please list examples.

Answer(s):
The most hazardous time we are most vulnerable to an incident is during transient operations.

Examples of transient operations include shutting down or starting the facility or equipment, taking equipment out of service, preparing to open the equipment for cleaning or internal maintenance, conducting maintenance onstream, running plant tests, or being in an operating upset condition.

10 Before opening equipment to the atmosphere, it must be ________, __________, ________, or chemically cleaned to remove the hydrocarbons and chemicals before air can enter it for the first time, possibly in several years.

Answer(s):
drained, purged, steamed

11 When preparing for equipment startup, essentially, all of the ________ must be removed from the equipment before _______________ are admitted into it, or the hydrocarbons can mix with the air, potentially creating an explosive mixture.

Answer(s):
air; hydrocarbons

12 The OSHA requirements for entering a confined space, such as a process vessel, a rail car, or a tank truck, are readily available in the Standard ________________.

a OSHA Standard 29 CFR 1910.147
b OSHA Standard 29 CFR 1910.119

- **c** OSHA Standard 29 CFR 1910.95
- **d** OSHA Standard 29 CFR 1910.146

Answer(s):

d OSHA Standard 29 CFR 1910.146

13 What is meant by the term life-saving rules?

Answer(s):

Life-saving rules are safety rules designed to protect against situations in which a single unsafe act, the failure to follow a critical procedure, or similar intentional errors can result in severe injury, significant loss of containment, or even loss of life.

14 What is the expected consequence of violating a life-saving rule?

Answer(s):

A violation of a life-saving rule can lead to the termination of employment.

15 Which of the following work rule violations should be considered a life-saving rule?

Please select the correct answers.

- **a** Failure to wear a seat belt while driving inside the facility.
- **b** Failure to wear fall protection when working on an incomplete scaffold.
- **c** Failure to arrive at work at the prescribed time.
- **d** Failure to properly utilize the work permit system when opening a flange on a pipeline.
- **e** None of the above.
- **f** All of the above.

Answer(s):

- **a** Failure to wear a seat belt while driving inside the facility.
- **b** Failure to wear fall protection when working on an incomplete scaffold.
- **d** Failure to properly utilize the work permit system when opening a flange on a pipeline.

16 Please describe in your own words what the term normalized deviation means.

Answer(s):

Normalized deviation describes a process where a work rule or procedure is violated by an individual or a workgroup, and there appears to be no consequence. This happens so often that it becomes the accepted way of performing a task. Eventually, the situation changes, and a catastrophic event occurs, where, if the task had been completed according to the work rules or procedure, the incident could have been avoided.

17 Following an actual event or even a near loss, the most important thing is ensuring that the generated ____________ are documented fully and ____________.

Answer(s):
recommendations; implemented

18 What are the common issues with preparing exchangers for blinding or opening when configured in parallel trains?
Please select the most correct answer.

- **a** It is difficult to verify that the product was completely drained from all exchangers.
- **b** Chemical cleaning may be ineffective due to fouling in some exchangers.
- **c** The exchangers may be physically located in a heavily congested work area.
- **d** One bank of the exchangers may be heavily fouled, and flushing may be ineffective, leaving hydrocarbons trapped in the exchangers.
- **e** None of the above.
- **f** All of the above.

Answer(s):
f All of the above.

19 First and foremost, we must recognize that we are vulnerable to the release of fluids or gases during all phases of the OPE process.

In our business, these fluids may be highly ____________ like petroleum naphtha or gasoline, they may be extremely volatile like ____________ ____________ ________ or may be highly toxic like ________ ________.

Answer(s):
flammable; liquefied petroleum gas; hydrogen sulfide

20 The Pemex refinery in Deer Park, Texas, experienced what appears to have been an OPE incident in October 2024 when contract personnel were in the process of ______________________________. This was a tragic incident where two contractors were killed, 13 more were injured, and a major release of hydrogen sulfide was released into the refinery and the surrounding communities.

Please select the most correct answer.

a Removing access plates from a process vessel.
b Removing a process pump to send to the shop.
c Opening a process exchanger for maintenance.
d Installing isolation blinds in a pipeline system.
e None of the above.
f All of the above.

Answer(s):
d Installing isolation blinds in a pipeline system.

21 What is a common hazard when opening a distillation column or a scrubber that has been in high hydrogen sulfide service?

Answer(s):
Opening distillation columns that have been in hydrogen sulfide service has a high probability of containing pyrophoric materials.

22 Why is pyrophoric material a concern? What can happen due to pyrophoric materials in a distillation column?

Answer(s):
When a distillation column is opened and contains a pyrophoric compound, when air enters the column, the pyrophoric material can ignite and ignite internal components in the column, such as internal demister screens or metal packing material. This can cause the column to collapse.

23 What are the different types of isolation blinds and how are they used when isolating process equipment?

Answer(s):
The three types of isolation blinds and their uses:

- A standard flat plate blind – used for regular isolation without concern about creating a trapped blind or the need for having an air gap or gas test location.
- A stand-off blind with stand-off tabs to provide separation – used to provide an air gap separation and to provide a good location for gas testing.
- A vented blind – used to prevent the installation of a trapped blind or location where the pressure of fluid cannot be vented or drained.

24 Cone roof design storage tanks are typically designed with a relatively ______ top roof ________.

With this design, if an internal explosion occurs inside the storage tank, the roof seam is designed to _________, releasing the pressure in the tank. This prevents the loss of containment that can occur when the _________ seam fails.

Answer(s):
weak; weld or seam; fail; bottom

25 Which of the following gas tests should be completed before entry into a permit-confined space?
Please select all that apply.

- **a** Oxygen content (%)
- **b** Hydrogen sulfide content (ppm)
- **c** Flammables (LEL)
- **d** Other toxic materials that may be associated with this process unit or area.
- **e** All of the above.
- **f** None of the above.

Answer(s):
e All of the above.

26 What are the acceptable limits set by the US Occupational Safety and Health Administration for the following?

Oxygen: Your Answer: 19.5% minimum

Flammables: Your Answer: 10% maximum

Hydrogen sulfide: Your Answer: 20 ppm maximum

End-of-Chapter Quiz (with Answers)

Chapter 1: End-of-Chapter Quiz (with Answers)

1 Which of the following are NOT considered opening process equipment?
Please choose all that apply.

- **a** Opening flanges on a pipeline to install isolation blinds.
- **b** Opening access plates on a storage tank to access the tank for cleaning and internal inspection.
- **c** Opening a suction strainer on a product pump to clean the strainer basket.
- **d** Opening an access place on a chemical railcar to gauge the level in the tank.

- **e** Opening the access plate on a product fractionator column for entry.
- **f** Draining water from a crude oil storage tank before lining up the tank to the crude unit.
- **g** All of the above.

Answer(s):

- **c** Opening a suction strainer on a product pump to clean the strainer basket.
- **d** Opening an access place on a chemical railcar to gauge the level in the tank.
- **f** Draining water from a crude oil storage tank before lining up the tank to the crude unit.

2 Opening equipment to the atmosphere creates the possibility of a loss of __________, such as a ________ or ________ of hydrocarbons, toxic chemicals, or a pressurized product. It also allows outside ________ to enter the equipment, which can react with flammable or __________ material.

Answer(s):
Containment; spill; release; air; pyrophoric

3 The facility is responsible for clearly defining the roles and responsibilities of all personnel involved in ________ work planning and equipment isolation and preparation for opening. All personnel must be ________ and ________ to safely perform and execute their assigned ________ before taking on this responsibility.

Answer(s):
OPE; trained; qualified; tasks

4 What are several potential advantages of a dedicated work permit for authorizing opening process equipment?
Please select all that apply.

- **a** A unique work permit dedicated to OPE would add additional emphasis to the hazards of OPE.
- **b** A unique work permit may help facilitate a checklist of potential OPE concerns to be addressed.
- **c** A unique work permit would facilitate more in-depth tick boxes, placing additional emphasis on other critical criteria that are more specific to OPE.
- **d** A unique work permit would set OPE work apart from other more routine cold work tasks, adding emphasis to OPE work.
- **e** All of the above.
- **f** None of the above.

Answer(s):
e All of the above.

5 Based on industry data, at what point do many of the loss-of-containment incidents often occur?

Please select all that apply.
a OPE incidents have occurred during in ser maintenance operations.
b OPE incidents have occurred during startup operations.
c OPE incidents have occurred during equipment blinding for isolation.
d OPE incidents have occurred during shutdown operations.
e All of the above.
f None of the above.

Answer(s):
e All of the above.

6 If a leak or release occurs while conducting OPE tasks for equipment isolation or entry, what should take place immediately?

Answer(s):
All OPE work will cease immediately, and flanges will be retightened if any leakage or spillage occurs.

7 Vents and drains are verified as free and clear and in a _______-________ state before a _______ _________ is issued.
If a _______ or ________ is fouled or plugged, a proper __________ ___________ device will be used to help verify that the bleeder is clear before proceeding.

Answer(s):
zero-energy; work permit; vent; drain; bleeder unplugging

8 What type of tag should be placed on all points of equipment blinding before the work permit for blinding is issued? When should this be installed at each blind point?

Answer(s):
A weather and tamper-resistant tag should be placed on each point of line or vessel opening before the work permit to install blinds is issued.

9 What special PPE is typically required if H_2S is suspected, or if it is identified during the preopening gas test?

Answer(s):
If H_2S is suspected or identified via the preopening gas test, breathing air (either a self-contained breathing apparatus (SCBA) or a hose line supplied) is required.

10 Why is it important to consider the potential of creating a "trapped blind" when installing blinds for equipment isolation?

Answer(s):
A trapped blind can accumulate process pressure or may contain hazardous materials and make it difficult to remove the blind.

11 How can the creation of a trapped blind be prevented?

Answer(s):
Installed a vented blind to allow venting of the trapped product to a safe location.

12 What is a special consideration when opening flanges on piping circuits or process equipment?
Please select the most correct answer.
a Consider the potential for creating a trapped blind.
b Ensure the correct bolts are being used.
c Ensure that all bolts are properly lubricated.
d Ensure the provision for collecting any trapped liquids.
e All of the above.
f None of the above.

Answer(s):
d Ensure the provision for collecting any trapped liquids.

Chapter 2: End-of-Chapter Quiz (with Answers)

1 What is generally required by a process facility to develop a procedure for safely conducting a unit shutdown and decontamination procedure (according to US OHSA)?
Please choose all that apply.
a The process and instrument drawings.
b The materials of construction.

c The process safety information.
d The properties of chemicals used in the process.
e The information on the unit's safety systems.
f The list of low-point drains and high-point vents.
g Equipment isolation and blinding points.
h The locations of steam tracing, etc.
i All of the above.
j None of the above.

Answer(s):
i All of the above.

2 The preparation and opening of refining or petrochemical plant piping and other associated equipment can be very ____________ and require specialized ____________, well-trained ____________, and specialized ____________ ____________ ____________.

Answer(s):
Hazardous; procedures; personnel; personal protective equipment

3 What is the primary difference between the preparation of a crude unit and a hydrotreater when preparing for isolation, opening, and entry?

Answer(s):
The crude unit is prepared by gas oil flushing, water washing, and steaming before setting blinds for opening and entry. In contrast, the hydrotreater is prepared by nitrogen purge to protect the catalyst in the reactor.

4 There are several general safety guidelines that should be followed when steaming process equipment when preparing for opening. Please name as many as possible.

Answer(s):

- Always be aware of the potential for steam burns or burns from hot condensate. All personnel involved in this activity should wear standard PPE such as gloves, face shields, rubber boots etc.
- Steam hoses should be identified as suitable for high-pressure steam use and properly rated for the temperature and pressure.
- Steam hoses should also be equipped with a whip check, a safety device installed between two steam hoses or between the manifold and hose to prevent personnel injury from a flailing hose.
- Quickly address any water hammering to prevent hose or piping failure.

- Steam should not be injected into low-point drains as this interferes with condensate draining.
- Ensure condensate is being continuously drained from all low points, including dead ends.
- Ensure that a slight positive pressure is maintained on the equipment to keep the temperature hot.
- Ensure system pressure is monitored continuously and is maintained below the MAWP. Continue steaming until a gas test proves the system is gas-free.

5 When are we most susceptible to finding pyrophoric material that can result in an internal fire when opening process equipment?
Please select the correct answer.
  a We are most susceptible to this when the process service is sour and contains a high concentration of hydrogen sulfide (H_2S).
  b We are most susceptible to this when the process vessel is corrosive and has deposits of corrosion products in the vessel's bottom.
  c Chlorine can result in pyrophoric, and therefore, vessels in chlorine service are very susceptible to pyrophoric materials.
  d Pyrophoric materials are not an issue at oil and gas facilities.

Answer(s):
  a We are most susceptible to this when the process service is sour and contains a high concentration of hydrogen sulfide (H_2S).

6 Please list the most important precautionary steps to take when opening a process vessel identified as possibly containing pyrophoric materials.

Answer(s):

- Always be aware of the potential for pyrophoric materials anytime equipment in sour service (H_2S) is being opened (such as storage tanks, fractionation columns, gas compressors, etc.).
- When opening equipment, place a charged fire hose strategically located at each manway being opened.
- Periodically spray water into the vessel manway and never allow the vessel interior to dry; otherwise, the pyrophoric material may ignite and ignite other flammable or combustible material. Water flooding of the column is not required; periodic wetting is needed to prevent the vessel from drying.
- During work preplanning, using a reputable chemical vendor, consideration should be given to pyrophoric neutralization of the vessel interior using a

weak solution of potassium permanganate. A reputable chemical vendor can help with this.

- In the work plan, we should expect to remove the demister screen and internal packing material from all areas where hot work is planned. Otherwise, they may ignite and create an internal fire that is extremely difficult to extinguish.

7 Why is it important when planning blinding locations to set vessel blinds at the first flange off the vessel on all lines?

Answer(s):
Blinding at the first flange off the vessel helps minimize interconnections with other equipment and the opportunity to miss a connection. It also helps eliminate dead legs or dead ends that may trap liquids or contaminate a vessel with hydrocarbons or toxics, possibly creating issues with a satisfactory gas test.

8 Before opening flanges for setting isolation blinds, whose responsibility is it to verify that all process equipment is properly drained and depressured and to ensure that the isolation valves are holding?
Please select the correct answer(s).

a Maintenance planners
b Maintenance craftsmen
c Area supervisors
d Operations

Answer(s):
d Operations

9 Why is it important to quickly install isolation blinds in fuel gas lines after shutting down a furnace, heater, or boiler?

Answer(s):
To prevent the possibility of an explosion, isolation blinds should always be installed in the main fuel piping circuits on furnaces, heaters, and boilers as soon as the equipment is shut down.

10 What makes flare systems unique when preparing for blinding or opening?

Answer(s):
Flare systems typically include very large diameter piping and are generally not designed to support the weight of water.

11 Why is it important to sequence the blinding of flare lines to prevent air entry into a flare line or flare system?

Answer(s):
Any air ingress can create a flammable mixture, resulting in an explosion in the flare system.

12 When blinding or opening piping or process systems that previously contained a hazardous chemical, operators should verify that the equipment and/or piping are completely ____________, vented of residual _________ (to a zero-________ state), and fully ____________.

Answer(s):
isolated; energy; energy; depressured

13 What hazard exists when purging process vessels and other equipment with nitrogen, and what special PPE may be required?

Answer(s):
Ensure precautions to prevent asphyxiation if nitrogen or another inert gas is used. This may include requiring a hose line supplied breathing air or self-contained breathing apparatus (SCBA).

14 What are the typical gas tests required before opening a process pipeline or vessel, and why are these performed?

Answer(s):
Before opening the process equipment, a gas test is performed to check for oxygen deficiency (O_2), flammables (LEL), and toxins (generally hydrogen sulfide [H_2S]). Tests for other toxins are based on the vessel's prior service.

This gas test helps determine the PPE requirements for those opening the process equipment.

15 When executing hot work in a process vessel and continuous gas testing is not done, a gas test should be completed within _______ hours or less before work begins, possibly sooner and with more frequent testing, depending upon the job and process conditions.

Answer(s):
2

16 When gas testing a process vessel that is at, or near, ambient temperature (below 100 °F or 38 °C), a gas test will not detect a ____________ ____________ that may be present in the vessel.

Answer(s):
combustible liquid

17 Why is washing or flushing through parallel trains such as parallel exchangers difficult?

Answer(s):
This happens most often due to internal fouling of the equipment or by the configuration of the parallel train. This can create internal pressure and flow restrictions and limit or possibly totally restrict the flow through one or more parallel trains.

18 Where should the fire watch be positioned when using a "fire watch" while hot work is in progress?

Answer(s):
The fire watch should be positioned away from the line of fire of the equipment being opened or the hot work but prepared for immediate response when needed.

19 The OSHA regulation requires that the fire watch remains on station for _____ minutes after the hot work is completed.

Answer(s):
30

20 Hot bolting replaces and tightens _________ bolt at a time without breaking the flange or seal.

Hot bolting should also require an _________ _________ to ensure that only one bolt at a time is removed, that it is promptly replaced with a fresh bolt, that the new bolt is fully torqued before moving to the next bolt for removal, and that the seal is never broken.

Answer(s):
one; independent observer

21 A special work permit is required to authorize entry into a designated confined space. Each entrant must be tracked, and vessel entry and exit times must be recorded for all entries.
True or False

Answer(s):
True

22 What are the US OSHA requirements for executing an entry into a permit-required confined space?

Please name as many as you can.

Answer(s):
Confined space entry requires special precautions, including the following:

- A satisfactory gas test before entry.
- Forced "clean air" ventilation for the duration that workers are in the confined space.
- A designated attendant to monitor the activities of the entrant(s).
- A personnel rescue plan.
- Warning signs to keep others away.
- If entrance covers are removed, the opening shall be promptly guarded by a railing, temporary cover, or other temporary barrier that will prevent an accidental fall through the opening and that will protect each employee working in the space from foreign objects entering the space.

Chapter 3: End-of-Chapter Quiz (with Answers)

1 A significant hazard associated with OPE is the _________ or _________ of process vents or drains. This may give the false impression that the system is at _______ pressure when fully pressurized.

Answer(s):
fouling; plugging; zero

2 In a loss of containment event that results in a large unit fire, the situation may rapidly escalate due to additional equipment damage and even more product losses. This may escalate to involve other process ________, resulting in _____________ damage or personnel __________.

Answer(s):
units; environmental; injuries

3 Another hazard associated with OPE is the fouling or plugging of process ________ or __________.

Answer(s):
vents; drains

4 What are the advantages of using a bleeder cleaner to unplug a vent or drain instead of a welding rod?
Please select all that apply.

- **a** The device is completely enclosed and protects workers from contact with the process fluid and a resulting injury.
- **b** It is designed with a pressure gauge to indicate when the plug has been removed.
- **c** It includes a vent that is used to vent pressure away safely or can be used to inject steam into the bleeder.
- **d** It comes in various design pressures and configurations, which should be chosen specifically for the piping system where it is to be used.
- **e** None of the above.
- **f** All of the above.

Answer(s):
f All of the above.

5 Simply stated, what is the normalization of deviance?
Please select all that apply.

- **a** A term used to describe an alternate process to the accepted process, which saves time when compared to the original process that the company specified.
- **b** A term used to describe a process where the organization requires an alternative process that is different from the approved process but has been approved by the team leader.
- **c** A term used to describe an alternative process that the team leader developed that saves both time and resources and is frequently used by all team members, although the organization does not require it.
- **d** A term used to describe a process where a person or an entire organization has developed a process that deviates from the established procedure and has used it to the point where it becomes the accepted new normal.

Answer(s):
d A term used to describe a process where a person or an entire organization has developed a process that deviates from the established procedure and has used it to the point where it becomes the accepted new normal.

6 When opening flanges or access plates on equipment or piping that has or may contain flammable products, the worksite should be surveyed to identify and eliminate any potential ________ ________.

Answer(s):
ignition sources

7 It is very important to ensure that all personnel understand that in an emergency or in the event any hazardous product is detected, all work should be stopped immediately and restarted only after the site is reevaluated and all clear is given. The work permit should be reissued before work restarts.
True or False

Answer(s):
True

8 What should be used to collect residual material when opening process flanges or access plates?

Answer(s):
splash pans

9 Always treat the opening of a flange to install or remove a slip blind, a figure 8 blind, or the removal of a blind flange as a line opening. They require a full ________ ________ and full line ________, including equipment ________ and ________.

Answer(s):
work permit; isolation; venting; draining

10 An accumulation of hydrocarbons may also result in a sudden and unexpected release and fire when equipment is being prepared for opening. Even what is normally considered a combustible, such as heavy distillate or diesel, can suddenly erupt in a dangerous fire, especially if any open flame or hot work is being done as part of the OPE.
True or False

Answer(s):
True

11 When process equipment is opened, there is the potential for release to the air, soil, and possibly to rivers, streams, lakes, etc. These may lead to ___________ damage and environmental _________ assessments, municipal _______ or penalties and damage to the company's ___________.

Answer(s):
environmental; damage; fines; reputation

12 When a release results in an explosion and unit fire, what should we be concerned about when additional equipment is affected?

Answer(s):
We must always be concerned about the potential for escalation to an even larger event. As additional equipment is affected, additional loss of containment of other process fluids and chemicals will inevitably occur.

13 Which of the following is the better and safer alternative for equipment isolation?
Please select all that apply.

- **a** Isolation using double block valves with a bleeder valve in between to verify no leakage.
- **b** Isolation with a single block valve with downstream bleeder valve open to verify no leakage.
- **c** Isolation with double block valves with bleeder valves upstream and downstream of the valves open to verify isolation.
- **d** Isolation using a single isolation valve close to the equipment and another isolation valve a long distance also closed to provide secondary isolation.

Answer(s):

- **a** Isolation using double block valves with a bleeder valve in between to verify no leakage.

14 One of the most significant hazards associated with OPE is the unexpected _________ of pressurized fluids or gases, including toxic gases.

Answer(s):
release

15 Releasing a hazardous product while opening process equipment can severely injure personnel and result in significant fire and property damage, especially if the released materials are _______ or _________.

Answer(s):
Toxic; flammable

16 Which two US Space Shuttles were lost due to normalization of deviance?

Answer(s):
Columbia; Challenger

17 Sometimes, equipment elevation can result in trapped liquids released during flange opening, allowing air to enter the system.
True or False

Answer(s):
True

18 Why should we NOT use a refinery air system for breathing air?

Answer(s):
There are many reasons why refinery air systems should not be used for breathing air. First and foremost, these systems are not properly tested and certified as breathing air.

The air may be contaminated, and the system may be backed up by nitrogen to protect the instrument air system from low pressure.

Some of these systems were labeled as "air" but, unfortunately, actually were nitrogen, and fatalities have resulted.

19 When purging equipment, why is it important to erect warning signs and barricades where nitrogen is venting, especially near vessel entry points like manways?

Answer(s):
To keep individuals who may not be familiar with the operations away. This is crucial near vessel entry points because nitrogen is odorless and colorless, and it displaces oxygen in the breathing zone. There have been instances where individuals have been overcome and fallen into a vessel undergoing a nitrogen purge, resulting in fatalities.

Chapter 4: End-of-Chapter Quiz (with Answers)

1 We must recognize that we are vulnerable to the release of toxic or flammable fluids or gases during all phases of the __________ process.

Answer(s):
OPE

2 Always remember that what has happened before can happen ________. This is extremely important, and we should recognize this is the first step in preventing it from happening at our facility.

Answer(s):
again

3 The Pemex refinery in Deer Park, Texas, experienced what appears to have been an OPE incident in October 2024 when contract personnel were in the process of installing isolation blinds in a pipeline system. This was a tragic incident where two contractors were killed, 13 more were injured, and a major release of ________________ was released into the refinery and the surrounding communities.
Please select the correct answer(s):

- **a** Hydrofluoric acid (HF)
- **b** Benzene (C_6H_6)
- **c** Hydrogen sulfide (H_2S)
- **d** Chlorine (Cl_2)
- **e** Monoethanolamine (MEA)

Answer(s):
- **c** Hydrogen sulfide

4 If the steam traps are discovered to be cold while draining fluids from low-point bleeders, liquids may be trapped in the system due to internal blockage caused by solidified products.
True or False

Answer(s):
True

5 What is unique about equipment like exchangers that are configured in a parallel flow path?
Please select the correct answer(s).

- **a** Following the process flow path through the exchangers is usually easy.
- **b** Parallel path equipment may contain trapped liquids that are difficult to clean and difficult to determine if they are present.
- **c** Trapped liquid hydrocarbons may be present even though a gas test reads 0% hydrocarbons.
- **d** Trapped hydrocarbons may be released when flanges are opened on exchangers configured in parallel paths.

e None of the above.
f All of the above.

Answer(s):
b Parallel path equipment may contain trapped liquids that are difficult to clean and difficult to determine if they are present.
c Trapped liquid hydrocarbons may be present even though a gas test reads 0% hydrocarbons.
d Trapped hydrocarbons may be released when flanges are opened on exchangers configured in parallel paths.

6 Following an actual event or even a near loss, the most important thing is ensuring that the recommendations generated are ___________ fully and ___________.

Answer(s):
documented; implemented

7 When gas testing a process vessel or exchanger at ambient temperature or slightly above after opening, what product or material will NOT provide a flammable reading on the gas test machine?
Please select the correct answer(s).
a Propylene at a temperature of 105 °F.
b Diesel with a flash point of 145 °F.
c Butane at a temperature of 65 °F.
d Gasoline with a flash point of −45 °F.

Answer(s):
b Diesel with a flash point of 145 °F.

8 Which of the following most correctly describes what is commonly referred to as "hot bolting"?
Please select the correct answer.
a Depressuring and clearing the system, then burning all the bolts except four, one at each quadrant, then burning those one at a time.
b Depressuring and clearing the system and heating the bolts with a torch before their removal.
c Depressuring and clearing the system, burning one bolt at a time, reinstalling the bolt with a new one, and tightening it before moving to the next one.
d Depressuring and clearing the system, performing a gas test to ensure that no hydrocarbons are present, and then burning all of the bolts.

Answer(s):

c Depressuring and clearing the system, burning one bolt at a time, reinstalling the bolt with a new one, and tightening it before moving to the next one.

9 If air is allowed to enter the equipment before preparation is completed, even if a very small amount of hydrocarbon or chemical remains, we can easily create an __________ __________ inside the process equipment being opened.

Answer(s):

explosive mixture

10 What could occur within the piping when draining a high-viscosity piping circuit for hot work, especially if the steam traps are malfunctioning?

Answer(s):

The material can solidify within the piping, causing blockages in the drains. Liquid hydrocarbons can build up in the piping and lead to a spill when the piping is opened or warmed, such as by ambient temperature.

11 Which of the following is likely while attempting to clear a bank of heat exchangers which are in series flow:

Please select the most correct answers.

a There is usually no problem clearing banks of exchangers that are in series flow.

b Series flow exchangers can create pressure drop causing low flow when washing the tubes.

c Sludge or coke material may accumulate in the exchanger tubes, causing significant amounts of liquid to accumulate in the exchangers.

d Significant amounts of liquid may build in the exchangers because of the pressure drop caused by series flow exchangers.

Answer(s):

c Sludge or coke material may accumulate in the exchanger tubes, causing significant amounts of liquid to accumulate in the exchangers.

12 When opening process equipment, what are some effective ways to prevent workers from opening the wrong flange or cutting into the wrong pipeline?

Please select the most correct answer(s).

a Ensure a good walk-down of the job site with the workers, discussing exactly what equipment will be opened before the work permits are issued.

b Ensure all flanges prepared for opening are properly tagged at the flange to be opened with a distinctive tag that is used only for this purpose.
c For piping in long pipe bands, ensure that only the piping to be opened is painted with a distinctive colored stripe and communicate to the workers that these are the only flanges to be opened.
d Ensure that the job site is visited frequently and verify that the status of the ongoing work is in compliance with the work permit.
e None of the above.
f All of the above.

Answer(s):
f All of the above.

Chapter 5: End-of-Chapter Quiz (with Answers)

1 The US Federal Occupational Safety and Health Administration (OSHA) has enacted and disbursed federal regulations for conducting process "opening process equipment"- type work into several existing federal regulations, rather than creating a special or dedicated regulation for this type of worker activity. For example, to cover the requirements for equipment isolation and preparation for inspection, testing, or maintenance, OHSA has enacted the federal requirements for energy isolation into the ______________________ standard affectionately referred to as the ________/________ standard, or 29 CFR 1910.147.

Answer(s):
"The Control of Hazardous Energy"
lockout/tagout

2 Many overseas and foreign governments have also enacted regulations that cover process safety specific to their ____________. While many essentially parallel the OSHA regulations, I am aware that some of the overseas agencies have requirements for incident reporting and prevention that ________ the US requirements.

Answer(s):
jurisdictions; exceed

3 Which OSHA Standard requires employers to develop, implement, and follow detailed work procedures and processes for handling or processing hazardous chemicals?

Answer(s):
OSHA 29 CFR 1910.119 "OSHA Process Safety Management"

4 Which OSHA Standard defines and establishes the requirements for a permit-required confined space and a non-permit-required confined space?

Answer(s):
OSHA 29 CFR 1910.146 "Permit-required Confined Spaces"

5 Which of the following are covered or required by the OSHA PSM Standard? Please select all that apply.
- **a** The PSM Standard specifies the categories of chemicals that are defined as toxic, reactive, flammable, or explosive.
- **b** The PSM Standard defines the training required by each employee presently involved in operating a process and for each employee before being involved in operating a newly assigned process.
- **c** The PSM Standards require the employer to inform contract employers of the known potential fire, explosion, or toxic release hazards related to their work and the process and explain the applicable provisions of the emergency action plan.
- **d** The PSM Standard requires the contract employer to prepare a record that contains the contract employee's identity, the date of training, and the means used to verify that the employee understood the training.
- **e** The PSM Standard requires the employer to develop and implement written operating procedures that provide clear instructions for safely conducting activities involved in each covered process, consistent with the process safety information.
- **f** The PSM Standard does not require any of this information.
- **g** The PSM Standard requires all this information.

Answer(s):
g The PSM Standard requires all this information.

6 OSHA standards for safe work practices and for specific equipment requirements are embedded in many OHSA standards. Please list as many other OSHA equipment-specific standards as you can.

Answer(s):
Examples of OSHA Standards that contain requirements for specific equipment.

Electrical safety	Hazard prevention and control
Emergency action plans	Injury and illness recordkeeping
Exit routes	Medical and first aid
Fire prevention	Personal protective equipment (PPE)
Fire safety	Respiratory protection
Hazard communication standard	Walking/working surface
Hazard identification and assessment	Worker participation

7 What sets apart an OSHA permit-required confined space from a non-permit-required confined space in the OSHA Confined Space Entry Standard?

Answer(s):
A permit-required space has the following characteristics:

- Contains or has the potential to contain a hazardous atmosphere;
- Contains a material that has the potential for engulfing an entrant;
- Has an internal configuration such that an entrant could be trapped or asphyxiated by inwardly converging walls or by a floor which slopes downward and tapers to a smaller cross-section; or
- Contains any other recognized serious safety or health hazard.

8 The permit-required confined space standard details other requirements that must be in place prior to entry and for the full duration of an entry. Please list as many of these additional requirements as possible.

Answer(s):
Also addressed are requirements and definitions for other safety precautions, including the following:

- The posting of "DANGER – PERMIT-REQUIRED CONFINED SPACE, DO NOT ENTER" signs,
- The requirement and duties of an attendant or "hole watch,"
- Space ventilation requirements,

- Pre-entry safety verification and entrant accountability,
- Protection against falling objects or personnel into the space,
- A rescue plan with personnel readily available,
- The requirements for an entry work permit process prior to entry,
- Other support gear like two-way radios, lighting, ladders, etc.

Chapter 6: End-of-Chapter Quiz (with Answers)

1 Gas testers must be ___________ trained on the specific equipment they will use at the facility. They must also pass testing and obtain ____________ from the facility before beginning gas testing in the field.

Answer(s):
trained certification

2 Before anyone enters confined spaces, such as excavations or pits deeper than ___________, they must be tested following the gas testing guidelines.

Please select the correct answer.
a 18 inches (1½ feet)
b 24 inches (2 feet)
c 48 inches (4 feet)
d 60 inches (5 feet)

Answer(s):
c 48 inches (4 feet)

3 All process vessels must be gas tested before they are entered, which includes the following.

Please select all that apply.
a Process fractionation column.
b Process chemical mixing drum.
c Excavation greater than 3 feet in depth.
d 45 K barrel storage tank.
e 6-foot tall 200-barrel open-top water tank.
f All of the above.

Answer(s):
a Process fractionation column,
b Process chemical mixing drum,
d 45 K barrel storage tank,
e 6-foot tall 200-barrel open-top water tank.

4 To ensure their qualifications remain current, certified gas testers require periodic refresher training at specified intervals, typically ________________.

Answer(s):
annually

5 Bump testing of gas testing devices involves using calibration gas to verify the testing device's ____________ and should be done ________.

Answer(s):
accuracy; daily

6 Many people have died when entering railcars and tank trucks because the atmospheres were not __________ before entry. Others also died while attempting to __________ the first victim.

Answer(s):
tested; rescue

7 OSHA standards define oxygen levels below ________% as oxygen-deficient and immediately dangerous to life or health, while atmospheres above __________% are considered oxygen-enriched and present a significant fire hazard.

Answer(s):
19.5; 23.5

8 When the gas tester is sampling and using a longer sample tube to reach the depths in a process vessel, they should be trained to allow enough time for the gas test device to draw the sample into the test equipment. They must keep the test tube in the testing area for ____________ ___________ to help ensure accurate test results.
Please select the correct answer.

a About 20 seconds
b No longer than one minute
c Five minutes
d Several seconds

Answer(s):
d Several seconds

9 For areas or tests without continuous monitoring, it is advisable to have the gas tester periodically return and ____________ the equipment to ensure no __________ gas has accumulated.

For example, it is good practice to perform return gas testing immediately after workers take a ________ and before they resume ________ after lunch, after any work stoppage, or if the confined space has been vacated for more than about ________ minutes.

Answer(s):
recheck; flammable; break; work; 30

10 What are the benefits of using continuous gas testing?
Please select the correct answers.

- **a** Continuous gas monitoring tests continuously for the presence and concentration of multiple gases, typically performing tests every few seconds.
- **b** Continuous gas monitoring provides audible and visual alarms when gas is detected above preset values.
- **c** Continuous gas monitoring does not depend on human intervention or testing.
- **d** The technology and availability of continuous gas monitoring equipment have greatly improved over the past few years.
- **e** All of the above

Answer(s):
e All of the above

11 Please describe the sequence that gas testers should use when conducting gas tests on process piping vessels, excavations, etc.

Answer(s):
The proper sequence for testing is to first measure the oxygen (%) content, then the hydrocarbon (LEL), and finally hydrogen sulfide (ppm). These basic tests are performed on almost all process vessels and excavations, followed by testing other chemicals specific to the unit or area involved.

12 When gas testing a floating roof storage tank for planned hot work, for example, welding inside the tank, why is it necessary to gas test the pontoons on the tank in addition to the interior of the tank?

Answer(s):
Flammable liquids may be trapped inside the pontoons due to corrosion and can enter the tank interior, leading to unexpected ignition, explosion, or fire.

13 If a gas tester performs a recheck while work is in progress and the gas test results indicate that the space is unsafe, what actions should the gas tester take?

Please select the correct answer.

a The gas tester should immediately notify their supervisor that the space is unsafe.

b The gas tester should caution the workers and inform them of the gas test results.

c The gas tester should call for a recheck of the space with a different gas test device.

d The gas tester should return to the operator shelter and conduct a bump test of his gas test device.

e The gas tester should immediately stop the work and evacuate the space, moving the workers to a safe area.

Answer(s):

e The gas tester should immediately stop the work and evacuate the space, moving the workers to a safe area.

14 Propane has a flammable range of between 2.2% and 9.6%. If the gas tester is sampling a propane vessel and there is a reading of 100% LEL, what is the minimum concentration of propane present?

Answer(s):

At least 2.2%

15 At what percentage LEL is there likely to be an explosion or fire if a source of ignition, such as an open flame, an arc, or a spark, is nearby?

Answer(s):

100%

Chapter 7: End-of-Chapter Quiz (with Answers)

1 At what point in the equipment isolation and preparation phase should an isolation blind list be created, thoroughly detailing each isolation blind point? Please select the correct answer.

a During the preplanning phase.

b After the unit shutdown, before field work starts.

c It's okay to make the list as blinds are being installed.

d As an integral part of the overall turnaround plan.

Answer(s):

a During the preplanning phase.

2 Please explain why three different types of equipment blinds are used and the primary difference between them.

Answer(s):
There are three types of blinds. One is a flat metal plate with a tab, and the second is a flat metal plate with a tab and spacers attached. The spacers provide an air gap for physical separation and to facilitate a gas test. The third type is a vented blind with a drain on one side, which is used to prevent the installation of a trapped blind.

3 Who participates in the joint jobsite visit, and what is its purpose?

Answer(s):
The joint job site visit is a pre-job walk-down of the job site between the permit issuer and the workers. The workers should point out every flange and/or access plate to be opened or removed.

This ascertains that there is no communication issue, and it verifies that the workers are familiar with the work to be completed.

4 Why must equipment isolation blinds be placed as near the isolated equipment as possible?

Answer(s):
Isolation blinds should be placed as near the isolated equipment as possible to minimize the number of connections and to prevent missing a connecting line that may still contain hazardous material.

5 What training should workers receive relative to opening flanges on process piping?

Answer(s):
Workers should be trained to never remove all the bolts or nuts from flange or access plate before attempting to break the seal. Instead, they should loosen the bolts and break the flange seating (the seal) while the bolts are loose but still in place. This allows the workers to quickly reclose the flange if pressure or fluids are discovered when the seal is broken.

Also, workers should be trained to take a position out of the line of fire while opening flanges.

6 What type of hot work is allowed in proximity to process equipment that is being opened for isolation blinding? Why is your answer important?

Answer(s):
No hot work, flame or non-flame, should be permitted in process areas where equipment and piping are being opened. This means no burning or welding, no arc gouging, and no gasoline or diesel-powered vehicles or engines of any kind.

This is important because these are ignition sources, and any released fuel may result in an explosion or major fire.

7 Many workers have been ________ by entering a confined space without following the confined space entry standard (OSHA 29 CFR 1910.146). It only takes ________ deviation or omission, and one or more people can die.

Answer(s):
killed; one

8 How long is the employer required to retain the completed permit-required entry work permit?

Answer(s):
The employer must retain the completed permit-required entry permit for one year.

9 The US OSHA Confined Space Entry Standard requires that before an employee enters the space, the internal atmosphere shall be ________, with a calibrated direct-reading instrument, for ________ content, for ________ gases and vapors, and for potential ________ air contaminants, in that order.

Answer(s):
tested; oxygen; flammable; toxic

10 Please describe what is defined as a "trapped" blind?

Answer(s):
A trapped blind is one installed without bleeders, vents, drains, or some other means of verifying zero energy on both sides of the blind.

11 What is the hazard with a trapped blind?

Answer(s):

Process fluids or pressure can be trapped behind the blind and may result in a release when removing the blind.

12 When preparing process equipment for isolation or opening, all forms of energy must be eliminated. The primary energy source is process pressure. What are other forms of energy that must also be eliminated?

Answer(s):
While not intended to be a complete list, other forms of energy are electrical energy, spring energy, radioactive energy, and gravitational energy. Of course, all forms of energy must be eliminated to protect workers.

13 Process personnel should verify that all pipelines and other process equipment to be opened are properly ________ and free of __________ material.

Answer(s):
drained; hazardous

14 This includes verifying that all drains, bleeders, and vents are fully open, free and clear. Ensure awareness that bleeders or vents may be ________; therefore, it is necessary to reaffirm that no plugging or obstructions exist.

Answer(s):
obstructed

15 Please remember that we should handle disconnecting a blind flange or a blank as a _______ __________.

Answer(s):
line opening

16 Is it acceptable to burn bolts to open process equipment?

Answer(s):
Hot work should never be used to burn bolts while opening flanges on piping or other process equipment.

17 Describe what is meant by "hot bolting."

Answer(s):
Hot bolting is described as burning one bolt at a time, replacing the bolt and tightening it before burning the next bolt. This process is continued until all the bolts are replaced, and then the flange is opened the normal way, without hot work.

18 What are the significant benefits of continuous gas testing compared to the daily gas test?

Answer(s):
Continuous gas testing (CGT), which leverages new technology, offers notable differences and benefits. CGT provides nearly continuous monitoring of multiple gases, including O_2 deficiency. One key advantage is that when daytime heating vaporizes flammable material from sludge or residues, this technology quickly alerts workers, preventing a potential explosion.

19 What is the purpose of the opening process equipment (OPE) tag, and how is it used?

Answer(s):
The operations team uses the OPE tag to identify flanges or access plates that have been prepared (energy-free) for opening. The OPE tags should be of a unique color and design and should only be used for this purpose, and no other purpose. They should be readily identifiable to the field personnel to indicate that the tagged flange or plate has been prepared and is ready to be opened for blinding or vessel opening.

20 Our experience shows that it is generally not wise to ________ a small section of the unit while the rest of the unit continues to operate. Operation risks require a thorough evaluation to assess the ________ and consider options such as postponing the work to a scheduled turnaround, a safer time when experienced crews are available, or when the inventory of hazardous materials is lower. An evaluation might reduce ________ by delaying work on piping systems still connected to equipment that is in operation.

Answer(s):
isolate; risk; risk

21 Following the guidance in this book, the OPE "Open Here" tag is installed on the specified pipeline and vessel flanges.
Please select all that apply.

- **a** There is no guidance for "Open Here" tags on pipeline flanges or vessel access plates.
- **b** Installing an "Open Here" tag on every other flange would be acceptable because workers can see them.
- **c** "Open Here" tags are used solely on pipeline flanges, not on process vessel access plates.
- **d** "Open Here" tags should be attached to each pipeline flange and process vessel access plate that will be opened or removed.

Answer(s):

d "Open Here" tags should be attached to each pipeline flange and process vessel access plate that will be opened or removed.

22 What is the purpose of the OPE "Open Here" tags, and why is this important?

Answer(s):

These tags are to prevent the workers from opening the wrong flange or access plate. This is important because lives have been lost and have been changed forever because workers opened the wrong flanges.

Chapter 8: End-of-Chapter Quiz (with Answers)

1 Please describe in your own words what the term "life-saving rules" means.

Answer(s):

These typically serve as safety rules designed to protect against situations in which a single unsafe act, the failure to follow a critical procedure, or similar intentional errors can result in severe injury, significant loss of containment, or even loss of life.

2 Regarding a life-saving rule, if one of these rules is violated, what is a possible consequence of the violation? Why should this be a possible consequence?

Answer(s):

Violation of a life-saving rule is a serious offense and may result in termination of employment. This consequence is necessary because violation of a life-saving rule can result in a serious injury or even death.

3 Full compliance with the life-saving rules should be considered __________ and a condition of __________.

Answer(s):

mandatory; employment

4 When working in high-concentration H_2S areas, putting on an H_2S detector should be like putting on your seat belt; you shouldn't have to think about it. True or False

Answer(s):

True

5 The practice of requiring life-saving rules has proven to reduce the number of ________ incidents, especially those that can result in, or have resulted in, catastrophic ________ or loss of _______.

Answer(s):
serious; injury; life

6 Please list as many of the potential life-saving rules as possible:

Answer(s):
List of potential life-saving rules (not intended to be a complete list and not in any particular order):

- Failure to use fall protection when working at unprotected heights. For example, on incomplete scaffolds, platforms, or uncaged ladders.
- Failure to obtain a hot work permit before burning or cutting.
- Entering a confined space entry without authorization.
- Disabling a safety-critical device without authorization.
- Failure to verify equipment isolation before beginning opening process equipment work.
- Failure to wear a seat belt when driving or riding in a vehicle.
- Alcohol or drugs found in the worker's system while working or driving.
- No smoking, except in clearly authorized designated areas.
- Always wear a personal H_2S detector when in H_2S areas.
- Never walk or place any part of the body under a suspended load.

Chapter 9: End-of-Chapter Quiz (with Answers)

1 Which of the following best describes "risk perception"?
Please select the most correct answer.
- **a** Risk perception is how we each judge the level of danger from a specific hazard.
- **b** Risk perception is how we respond to risks we identify while working.
- **c** Risk perception is how our team works together to mitigate or eliminate risks.
- **d** Risk perception is the unsafe acts I see others doing while at work.

Answer(s):
a Risk perception is how we each judge the level of danger from a specific hazard.

2 Please describe in your own words what the term "risk habituation" means.

Answer(s):

a Risk habituation refers to the risks we assume at work when things aren't going well.

b Risk habituation refers to the risks we assume when our leadership is absent.

c Risk habituation is a term describing how we become accustomed to risks through repeated exposure while working in dangerous environments.

d Risk habituation is a process used to help us identify and evaluate risks.

Answer(s):

c Risk habituation is a term describing how we become accustomed to risks through repeated exposure while working in dangerous environments.

3 Please describe the difference between risk perception and risk habituation.

Answer(s):

Risk perception refers to how each person assesses a specific risk, while risk habituation is the decrease in our sensitivity to that risk.

4 Studies show that risk habituation can be improved through bulletins, __________ alerts, periodic reviews of industry __________ and similar measures where procedures or shortcuts caused ____________.

Answer(s):

safety; incidents; accidents

5 Studies also emphasize the importance of ___________. Training should address the hazards of risk habituation by explaining that it is an issue and how it ___________, to boost workers' ____________.

Answer(s):

training; develops; awareness

6 For operations groups, the US Occupational Safety and Health Administration Process Safety Standard (OSHA) 29 CFR 1910.119 requires a review of operating procedures every ________ __________.

Answer(s):

three years

7 Please list the three industries or areas mentioned in this Chapter where risk habituation has occurred with frequency.

Answer(s):
Road maintenance, Use of escalators, Refining and petrochemical plants

8 Risk habituation happens when the ________ gets used to repeated __________ to risks, especially during daily tasks repeated multiple times without noticeable or felt consequences. When no one gets ________ or __________, the person's perception of risk decreases.

Answer(s):
Brain; exposures; hurt; injured

9 Risk habituation has been described as a dangerous survival mechanism that causes workers to take risks they normally wouldn't take. The problem is that the threat stays the same, but our reaction to it changes, making risk habituation a concern.
Please select true or false to this statement.
True
False

Answer(s):
True

Chapter 10: End-of-Chapter Quiz (with Answers)

1 Please describe what "normalized deviation" means in your own words.

Answer(s):
Deviations from safety systems, standards, procedures, or established guidelines are quickly accepted because there are no obvious negative consequences, and no one calls attention to the deviation or the bypassed system or process.

However, at some point, there may be a slight difference in the scenario where something as simple as a different weather situation, another piece of equipment already out of service, or something similar can lead to a catastrophic outcome.

2 At BP Texas City, the operators were reportedly deviating from the operation procedure routinely. Understanding the description of normalized deviation,

could this have been occurring at that site, and could it have contributed to this catastrophic process safety incident?

Answer(s):
Yes – all indications are that this was exactly what was happening at BP Texas City. This is reinforced by the investigation by the US Chemical Safety and Hazard Investigation Board (CSB). They found that no one had called attention to the existing procedures, even though they had been deviated from during most of the previous unit startups.

3 How is normalized deviation reinforced?

Answer(s):
One person in the organization might deviate from a procedure or take a shortcut, and the benefits are quickly seen as positive and reinforced by others. Others then start following the same shortcuts, and it quickly becomes accepted as the 'new way' of getting things done.

4 You are aware that on January 28, 1986, the Shuttle Challenger exploded 73 seconds into its launch, killing all 7 crew members.
What role did normalized deviation plan in this tragic accident?

Answer(s):
The subsequent NASA investigation concluded that a solid rocket booster (SRB) O-ring seal in one of the booster joints failed, causing a fire that allowed flames to reach the external fuel tank. The joint had leaked multiple times during previous launches, and because no failure had occurred before, it was considered an accepted risk.

5 Seventeen years after the loss of the Challenger, on February 1, 2003, the $4 billion Columbia was also lost, and the entire crew was killed. This incident resulted in NASA grounding the entire shuttle fleet for 2½ years.
Do you remember that normalized deviation played a role in this tragedy?

Answer(s):
The space shuttle Columbia disintegrated while re-entering Earth's atmosphere destroying the spacecraft and with the loss of all seven crew members.

The investigation revealed that during the launch on its 28th mission to space, a piece of foam insulation detached from the external tank, striking the shuttle's left wing.

A similar loss of foam insulation had also occurred during previous shuttle launches, some with nearly catastrophic consequences. However, even with the prior incidents, the shuttle was allowed to continue to fly.

6 How can normalized deviation be recognized and challenged in your organization?
Please provide examples of how this can be done.

Suggested response:

- Once identified, a deviation should never persist – it must be corrected or formally reviewed and approved.
- Discuss with the individual how the deviation can be reviewed and approved through the MOC process.
- Designate a key team member as the team champion for the initiative or system.
- Ensure your organization and team have a strong safety culture.

This culture should support and encourage reporting deviations and emphasize the importance of following standard operating procedures or practices, even when there are pressures.

Appendix C

Key Elements for Removing Equipment from Service and Preparation for Opening: A Job Aid for Removing Equipment from Service

Appendix C is not intended to be a procedure; however, it does contain many of the key elements that should be included in equipment-specific procedures developed by and for your facility. With permission from Wiley Publishing Company, a small portion of the following has been reused from Darryl's first Wiley Process Safety book, Process Safety, The What, Why, and How Behind Safe Petrochemical Plant Operations.

A key part of safe equipment isolation and opening is communicating with the personnel in the field who will perform the work. Communication ensures they clearly understand exactly which equipment to open, including specific details such as which flanges or access plates will be opened. It also confirms that the equipment has been made safe for them to open and highlights hazards they should be aware of during the work. This approach means we do not rely solely on verbal instructions or explanations but also provide visual cues for workers to follow in the field. Visual cues like identification tags and paint markings on the piping to be cut or opened help ensure they don't open the wrong flanges or cut into the wrong pipe.

You may ask why this is necessary. Several recent incidents have involved workers in a congested piperack opening flanges to install figure-8 blinds, inadvertently moving onto the wrong pipe, and opening a flange on an in-service pipeline. These events have resulted in a catastrophic loss of hydrocarbon containment, leading to major refinery fires. In one case, a pipeline was opened, releasing hydrogen sulfide at a concentration of more than 90% (>900,000 ppm H_2S). This caused two workers to die and forced major metropolitan areas to shelter in place.

This is why we need to take extra time and effort to tag each point where the process will be opened for blind installation or equipment access for any reason. The protocol below describes several types of tags used for this purpose. If your site does not currently follow this protocol, it is strongly recommended to implement it before the next planned downtime. Lives could be saved by doing so.

Equipment Isolation

Equipment Opening Identification Tags ("Open Here Tag")

Equipment opening identification tags, the "Open Here Tag," can help prevent opening the wrong flange or a flange or access plate that has not been prepared. The permit issuer and the craftsmen should discuss the tags used to identify the equipment to be opened to ensure a clear understanding of this process. This is typically done during the pre-job meetings and field walkdowns.

Equipment opening identification tags (Open Here Tags) should be designed and attached to each flange that will be opened. The Open Here Tags should be unique in design and background color to distinguish them from other equipment identification tags.

The operations group should attach a unique Open Here Tag to each flange that will be opened only AFTER the equipment has been cleared and is ready for opening, but BEFORE any work permits are issued to open flanges or access plates. The attachment method should be sturdy and weather-resistant.

For planned piping cuts, the cut points should be clearly marked on the pipe, and an Open Here Tag should be placed next to the planned cut. These details should also be discussed with the craftsmen during the pre-job meetings and field walkdowns.

Specifically, the work crew must understand that they should never open flanges, access plates, or cut any piping unless they are clearly marked with this unique "Open Here" tag.

"Do Not Open" Vessel or Piping Isolation Valve Tags

The valves used to isolate a process vessel or a piping circuit should be selected in advance and located as close to the vessel as possible. Locating the isolation valve close to the vessel helps eliminate sections of piping that may contain hydrocarbons and prevents missing a pipe intersection or small piping during pre-job walkdowns or piping isolation.

A "Do Not Open" valve tag should be created to identify the preselected isolation points where valves or isolation blinds will be employed to securely isolate a process vessel or piping circuit for opening process equipment. The "Do Not Open" valve tag should be securely attached to each pre-identified isolation valve or blind point location before a work permit is issued to open flanges or access plates.

The isolation valves should be marked on the process P&ID drawings and in the field, and the distinctive isolation "Do Not Open" valve tag should be attached to the isolation valve or the blind flange if it is used for equipment isolation.

Zero-energy Verification Tag

Zero-energy points shall be verified before work permits are issued to open flanges, access plates, or install isolation blinds. These zero-energy points are typically process bleeders, such as those between double block and bleed (DBB) isolation points, but they may also be bleeders on other process equipment, such as pumps, exchangers, or piping circuits with single isolation points. A tag shall be developed to identify bleeders or other points used in the field where zero energy has been verified.

After field zero-energy verification, the operations group will attach this zero-energy verification tag at the verification point. The work permit issuer and the craftsmen will discuss these tags during the pre-job meeting and field walk-down. The objective is that the craftsmen understand how operations verified zero energy in the equipment.

The zero-energy verification point may be an electrical circuit breaker for motor-driven pumps or other electrical equipment. In some rare cases, spring energy may have been suppressed; nevertheless, a zero-energy verification tag should be used to identify how zero energy has been confirmed.

Joint Pre-job Meeting and Field Walkdown

During the pre-job meeting, the work permit issuer and the craftsmen will discuss the difference between the three tags: the "Do Not Open" valve tags used for vessel or piping isolation valves, the "Open Here Tag" used to identify equipment ready for opening, and the "zero-energy" verification tags.

It is critical that the workers clearly understand that they shall NOT open flanges that are not properly tagged with the "Open Here" tags, indicating that the equipment is ready for opening.

Energy Isolation

An energy isolation circuit should be established with isolation points as close to the flange or equipment to be opened as possible. Ideally, DBB should be used, with the bleed valve positioned between the double block valves. When the double block valves are closed and the bleeder valve is open without leaking products on both sides of the opening point, no energy should be at the opening point.

If hydrocarbon or other chemicals leak from the bleeder valve, it indicates that zero energy has not been achieved and could lead to a loss of containment. This should trigger a risk assessment and the development of a mitigation plan for how to proceed. A higher level of approval should also be required before moving forward with the detailed mitigation plan.

If a DBB valve arrangement cannot be used and the opening will be against a pressurized single block valve, the pressure on the upstream side of the isolation valve should be reduced to as low as possible. Opening a process pipe that has only a single isolation valve on each side is a practice that carries inherent risks, especially if the valves are potentially leaking internally. While DBB or isolation with a blind flange are preferred methods for ensuring positive isolation during maintenance, there may be instances where you have to work with single isolation valves.

Working against a single isolation valve, especially in older equipment, requires additional steps to guard against a potential loss of containment. The following offers suggestions for an action plan for opening a process pipe circuit with single isolation valves, with worker safety as a top priority.

Please recognize that even with these additional steps, when working against a single isolation valve, there is a risk of loss of containment:

- **Preliminary Hazard Analysis**
  - **Hazard Identification and Risk Assessment:** Review the contents of the piping and evaluate the inherent hazards of the product (consider flammability, toxicity, light products such as liquid petroleum gas (LPG), and potential pressure of the contents.
  - **Review Isolation Procedures:** If DBB isolation is not possible, follow procedures and site safety standards and policies using single valve isolation.
  - **Isolate and Depressure the System:** Block in the piping system and ensure all energy sources are isolated and depressured. The lockout/tagout (LOTO) process should be followed to ensure complete energy isolation.
  - **Drain and Purge the System:** Drain, depressurize, and/or purge the piping to remove hazardous materials as much as possible, following standard operating procedures.
  - **Be Aware of Potential High Temperature:** If necessary, provide cooling of the piping system by using fire water sprays or other similar methods to cool the piping to a safe working temperature. If the system is equipped with steam or electrical heat tracing, these should be turned off or removed from service.
  - **Consider Fire Protection:** Consider having additional fire protection equipment on standby, out of the line of fire but ready to respond if needed.

- **Single Valve Isolation and Verification**
  - **Close and Secure the Single Isolation Valves:** Close the single isolation valves on each side of the piping section to be opened and follow the site requirements for tagging the valves.
  - **Verify Isolation:** Verify isolation using the existing single isolation bleeder valve(s). In the event that one or more of the bleeder valves appear to be fouled or obstructed, use a fully enclosed bleeder cleaner tool to verify that the bleeder is unobstructed.
  - **If Necessary, Isolate the Adjacent Systems:** If the single isolation valves are leaking and the system remains pressurized, isolate and depressurize the adjacent piping systems.
  - **Lock LOTO:** Implement LOTO on the closed valves to prevent accidental operation. Ensure that lockout devices are applied to the single isolation valves and that tags are in place to indicate the purpose of the lockout and other pertinent information.
  - **Check for Leaks:** Verify that the single valve is not leaking through. This may involve checking a bleeder valve downstream of the isolation or using a pressure gauge to monitor for pressure build-up.
  - **Mechanically Lock or Secure the Isolation Valves:** Lock the valve wheels or valve handles to prevent the valve from being opened during the period the piping is open. If the system is provided with automatic valves, ensure that the actuators are locked.
- **Preliminary Gas Test Before Opening Flanges**
  - A preliminary gas test should be conducted using bleeders or other means to determine the atmosphere inside the equipment before opening flanges.
  - The primary purpose of this test should be to help determine the appropriate personal protective equipment (PPE) for the workers.
  - The test should include testing for oxygen content, hydrogen sulfide, and lower explosive level (LEL or explosivity) at a minimum. Consideration should be given to other chemicals, respiratory, or other physical hazards that may be present based on the chemicals used in the process.
- **Line Opening and Maintenance**
  - **Wear Appropriate PPE:** Ensure all personnel involved wear the correct PPE, including impact-resistant gloves, face shields, or goggles. In some cases, where H_2S may be present, workers may require a self-contained breathing apparatus (SCBA).
  - **Controlled Opening:** When breaking containment, consider opening a bleeder valve, a fitting, or a flange slowly to vent any residual pressure.

- **Monitor and Reevaluate:** Periodically monitor the isolation and reevaluate the risks throughout the task. This is especially important if the task is long-term.
- **Minimize Work Time:** To limit the potential exposure to personnel, the working time should be minimized to the extent possible.

Caution About Material Trapped Under a Valve Gate

When closing a rising stem valve for equipment isolation, verify that no more than a quarter to half inch of the valve stem remains above the valve wheel. A valve stem above this may indicate that debris or sludge is trapped beneath the valve seat, which could cause the valve to leak and potentially result in loss of containment when opening a downstream flange.

Clearing Equipment

When a process unit or even a small segment of process equipment is shut down for maintenance or a planned turnaround, hydrocarbons and chemicals must be removed from the equipment, and it must be vented to the atmosphere. The hazard of doing this is that if we allow hydrocarbons or chemicals to mix with air, it can cause a massive explosion or a major fire inside the process equipment when fuel and air combine. Therefore, we need to develop step-by-step procedures for safely removing hydrocarbons and other chemicals from the equipment and for safely venting the equipment to the air.

This process is quite straightforward for crude units and some other process units: first, we flush the unit with a midrange hydrocarbon, such as gas oil, which has a slightly higher consistency than kerosene. It is a relatively safe fluid because it has a high flash point and a midrange boiling point, making it effective for flushing. After this, we perform a water wash by circulating water through the unit to remove any remaining hydrocarbons or sludge. Next, we introduce high-pressure steam to the entire unit, first directing it to the flare system, then releasing it to the atmosphere. Once this is complete, we are ready to install the isolation blinds to segregate the shutdown equipment from the other process equipment.

For other process units such as hydrofiners (also known as hydrotreaters or HDS units), hydrocrackers, reformers (sometimes called platformers or powerformers), fluid catalytic cracking units (also referred to as cat units or cat crackers), alkylation units (also called alky units), various kinds of light ends units or polymerization units, and others, we need detailed procedures to properly and safely shut down and clear the units. This is because some of these units contain very expensive catalysts that could be damaged by exposure to steam or moisture. Some have alloy metallurgy that may crack if exposed to water containing chlorides. Additionally, some process equipment, like flare systems, is not designed to support the weight of water. The key point is that all these specifics must be considered when developing shutdown and clearing procedures, so each process unit will have clearly defined procedures specifically created for that unit or process equipment.

For example, when cleaning a hydrotreater, we typically use pressurized nitrogen to purge the unit. These units contain a large amount of stainless steel that could crack or be damaged by even small amounts of chloride in wash water, so any water, if used, must be tested in advance to ensure that it is essentially chloride-free. With pressurized nitrogen, we usually perform concertina purging or pressure and depressurization cycles to remove any remaining hydrocarbons from the piping systems. During this process, the reactors are isolated under a nitrogen blanket and maintained at a slight pressure to protect the catalyst. Flare systems also use nitrogen to purge hydrocarbons because most of these systems involve large-diameter piping that is not designed to support the weight of water.

Equipment Blinding

There should be a pre-job checklist, and a pre-job field visit between the work permit issuer and either the craftsmen or their representative, who is expected to convey the discussion that occurred during the pre-job visit to each of the craftsmen.

This checklist should cover the following as a minimum:

- Describe the tags and labels used at the facility. Include a warning that flanges or vessels must not be opened unless specifically tagged as "Open Here."
- Explain how energy isolation was accomplished in this task.
- Verify that the craftsmen understand the types of isolation blinds, their purpose, and examples of where they are used.
- Describe the emergency egress and evacuation routes at the worksite.
- Describe the fire and other emergency response requirements, including safety watch protocols.
- Discuss the gas test that represents the contents in the piping system or vessel to be opened.
- Review the LOTO procedures currently in place to ensure the safety of personnel.

- Review the necessary PPE that must be worn during all phases of the task.
  - Consider SCBA, especially if there's potential exposure to H_2S, nitrogen, or other toxic or asphyxiating substances.
  - Consider wearing a face shield, goggles, heavy rain gear, etc.
- Review the required barricades and warning signs.
- Discuss precautions if pyrophoric materials may be encountered, especially if the equipment being opened has processed even a small amount of hydrogen sulfide (see additional details below).

Pyrophoric Materials

All vessels handling or processing hydrogen sulfide, even at low concentrations and in low-oxygen environments where iron or steel is present, can produce significant amounts of pyrophoric materials. The pyrophorics form through a chemical reaction between sulfur compounds and iron or steel, creating iron sulfide scale, which is highly pyrophoric. This is especially true when these materials are in contact for an extended period. This situation applies to many refining processes, with the possible exception of products downstream of hydrofining processes, where the sulfur compounds have been removed.

We should expect to find pyrophorics in crude tanks, crude piping systems, and processing units. Pyrophorics are also present in hydrofiners, hydrotreaters, and hydrocrackers, including in the reactors, fractionators, MEA treaters, and sour naphtha and asphalt tanks. It is common to find pyrophoric compounds in exchanger tube bundles and pipelines, especially when these are in sour service with H_2S present. Other process equipment, such as internal demister screens and internal random or structured packing, may also contain pyrophoric materials that can ignite in the presence of air and easily spark other combustible materials. This is especially true if H_2S is present at nearly any concentration during processing.

The risk of pyrophoric materials is that they can ignite spontaneously when exposed to air, such as when opening an exchanger or a fractionator. Materials such as iron sulfide scale (metal rust with sulfur compounds) ignite easily upon contact with air. However, the real danger is that they can ignite other combustible substances they come into contact with, such as hydrocarbons or sludge. Even more concerning, a pyrophoric fire can ignite internal process vessel components. For example, many internal fires in fractionators start when a pyrophoric fire ignites internal demister screens or packing materials. When this occurs, the fire burns intensely and is difficult to extinguish. Usually, in such cases, the vessel's shell weakens due to the internal fire, leading to the vessel's collapse. Of course, the collapse of a major process vessel inside a refinery can cause additional releases of containment, a rapidly spreading fire or explosion, and catastrophic damage, with the potential to result in loss of life.

Mitigating Pyrophoric Hazards

A practical approach is to ensure the equipment is water-wet when exposed to the atmosphere. Water flooding is usually unnecessary; however, periodic water spraying should be considered to prevent pyrophoric materials from drying out.

If work is scheduled near the demisters or internal packing, it is preferable to remove these materials from the fractionator or ensure periodic water spraying to prevent them from drying out. Another effective practice is to position a charged fire hose at each manway where there is a potential for ignition by pyrophoric materials, with someone responsible for periodically walking up and down the tower to spray water into each manway. The key to preventing ignition is maintaining high humidity inside the vessel with regular water sprays. Early fire detection can be achieved by continuously monitoring the temperature and carbon monoxide levels in the column while it is open to the atmosphere. Due to some internal fires occurring almost instantly when the access plates were removed, it is a good idea to have the fire protection in place BEFORE opening the access plates.

If hot work is to be done inside the tower and near the packed bed, it may be wise to include removing the packing in the work plan. Otherwise, plan to install internal barriers to protect the packing from accidental ignition caused by hot work. Metal fires have occurred due to spontaneous combustion from increasing the temperature inside the tower; therefore, if the packing is not removed, periodic wetting must continue throughout the maintenance operation.

Chemical Neutralization of Pyrophoric Compounds

An effective method involves chemical neutralization before opening the equipment. During the final water wash step, this is usually done with a 1% potassium permanganate solution. Potassium permanganate is reactive; make sure to work with a reputable chemical vendor and an experienced chemical engineer, and thoroughly review hazards before use. Do not use potassium permanganate in acid service or on equipment containing low pH materials (refer to MSDS and chemical vendor).

The chemical vendors will circulate and monitor the color of the potassium permanganate solution (purple). Neutralization can also be performed after chemical cleaning or de-oiling (e.g., Zyme-Flow). However, chemical cleaning or de-oiling alone has proven to be ineffective as a neutralization method for pyrophoric compounds. This is especially true for pyrophoric materials that are deeply embedded in demister screens or towers with internal packing. If neutralization is not performed, periodic water sprays will be necessary to prevent the risk of pyrophoric ignition. Do not allow pyrophoric materials to dry in the presence of air.

Reactors and Catalysts

For reactors and catalysts where circulating the chemical solution or water wetting isn't possible, use nitrogen or other inert gases to prevent oxygen contact. Keep awareness of nitrogen hazards (such as low oxygen environments) and manage them accordingly. Tanks, reactors, columns, and exchangers in high-sulfur service must be properly blanketed with N_2 during periods of inactivity.

The catalyst drums should also be inerted to displace air during the loading of spent catalyst. A common practice is to purge the drum with N_2 and then insert dry ice (CO_2) into it before sealing to prevent oxidation of the pyrophoric material.

Sludge and Scale

Enforce the requirement that all deposits (scale/sludge, etc.) must be removed from the vessel immediately after opening. Keep all materials wet until they are taken out of the vessel and placed in a safe location. Remove sludge, debris, old demister screens, and old packing materials to a safe area away from the unit (to prevent combustion if they dry). Make sure that personnel handling the removal are aware that the material may be pyrophoric and understand how to handle it properly (keep it wet!).

Pyrophorics/Our History

At some point, most petrochemical sites have experienced the spontaneous ignition of iron sulfide on the ground or inside equipment, sometimes with disastrous results. Make sure mitigation procedures are included in the planning process and strictly followed when opening equipment. Keep deposits that may be pyrophoric wet until they are removed from the vessel and the unit.

Appendix D

The Importance of Effective Equipment Isolation: A Job Aid for Effective Equipment Isolation

This is not a checklist; however, the following are recommended items to include in a checklist for preparing equipment before and during the actual opening of process equipment.

Hazards associated with chlorine (Cl_2), hydrogen sulfide (H_2S), hydrochloric acid (HF), ammonia (NH_3), etc., are all different. This is why equipment preparation checklists should be tailored to the site's facilities, chemical hazards, and equipment.

The following information is not meant to be comprehensive and should not be used in the field for equipment isolation. This should be adapted for the facility involved.

Creating the Isolation Plan

The operations team is responsible for isolating process equipment. They must ensure that all equipment is disconnected from live or operational systems and all energy sources. All energy must be fully isolated before work permits are issued for opening pipe fittings, connections, flanges, access plates, and other openings into process equipment. Similarly, for electrically powered, steam-powered, or equipment and devices powered by alternative energy sources, all energy must be isolated before issuing a work permit to perform maintenance. For electrically operated equipment, it is strongly recommended to attempt starting the equipment at each starting location after isolation is complete. This helps verify that all energy has been properly isolated.

In rare cases where energy cannot be completely isolated, such as when isolation valves leak, the operations team must ensure that a formal risk assessment is completed, and alternative mitigation measures are in place before attempting to open equipment. Mitigation should eliminate hazards caused by the leaking valve. A higher authority must review and approve the risk assessment and mitigation plan before issuing the work permit.

Before isolating process equipment, the operations team should develop an isolation plan. The P&ID should be used as a reference, along with a walkthrough of the unit, to identify all isolation points, high point vents, and drains (bleeders). When creating the isolation plan, the isolation valves closest to the equipment that will be opened should typically serve as the isolation points. This helps ensure no branch connections are overlooked and reduces the amount of products or chemicals that may become trapped in the piping.

The isolation valves should be labeled on the P&ID and in the field, and they should have isolation tags that identify the points used to isolate the equipment. These isolation tags should be unique labels created specifically for this purpose and must be clearly visible to all personnel in the field.

The equipment isolation plan should specify the types of blinds to be used and their installation locations. Proper planning is crucial to prevent trapped blinds and to ensure stand-off blinds are employed for air circulation and gas testing points on isolated vessels and piping when hot work is scheduled. The types of blinds, piping blanks, and their applications include the following:

- Flat plate isolation blinds, with a tab, sometimes called a spade blind;
- Figure 8 blinds, also known as spectacle blinds;
- Stand-off blinds with tabs attached to one side to provide air circulation and gas testing;
- Bleeder blinds, sometimes known as vented blinds (with one side vented to a bleeder valve to prevent a "trapped" blind);
- Piping blank flange, including blank flange with bleeder valves attached.

Please refer to Chapter 7 for additional discussion on types of blinds and photos of each type.

High-point vents and drains are used to clear equipment of products or chemicals before issuing a work permit to open process equipment. As noted earlier, these should be identified as part of the isolation plan, and unique tags should be prepared in advance for each high-point vent and low-point drain (bleeder). These tags will be installed when the vents and drains are opened and will remain in place while they are open. Again, these tags should be distinctive and clearly visible to all personnel in the unit while installed. This helps ensure everyone is aware when the vents and drains are open and confirms they are properly closed before returning the equipment to service.

As I have emphasized many times in this book, make sure that identification tags are used to clearly mark all flanges and access plates approved for opening. Also, ensure that all workers understand they are not allowed to open any flange or remove any bolts on equipment that has not been tagged for opening.

Clearing the Equipment for Opening

When preparing equipment for opening, it is important to remove the product or chemicals from the piping, pumps, and other equipment beforehand. Each facility should create a plan tailored to its specific equipment, chemicals involved, and potential hazards. This plan can serve as a guide for developing checklists or procedures; however, it should be seen as general guidance rather than a set of detailed instructions and, therefore, not used directly in the field.

A word of caution: When isolating heat exchangers or similar equipment, always start by isolating the hot side. During recommissioning, ensure the cold side is commissioned before opening the hot side. This prevents the thermal expansion of trapped liquids on the cold side, which could cause a rupture. When developing checklists, be sure to include this caution.

Generally, when preparing equipment, it may be filled with a high-flash product like gas oil or possibly a kerosene-based product. This flushing is done either through existing piping to a slop tank or by using temporary hose connections. For lighter hydrocarbons like LPGs, the equipment may be isolated and then vented to the flare system for depressurization.

After the initial flush, the equipment can be rinsed with water to remove the distillate and direct it to a safe location such as a slop tank. Water is then drained from the low-point drains, and the high-point vents are opened to prevent creating a vacuum in the equipment. Afterwards, many systems use high-pressure steam to eliminate any remaining hydrocarbons. A temporary steam hose is attached to the equipment, allowing operators to steam the equipment either to a unit blowdown system or possibly to the atmosphere.

Appendix E

Field Checklist for Equipment Opening: A Job Aid for Safe Equipment Opening

As you are aware from the incident cases in this book, we have experienced several catastrophic incidents caused by workers opening the wrong flange and inadequate preparation of equipment for opening. I have detailed procedures and practices in this book to help prevent additional incidents of this kind; however, I understand this involves using several tags and other techniques that may be confusing.

What follows is a checklist designed to simplify these techniques. This isn't a ready-made checklist, but it can be helpful as each facility creates procedures or checklists tailored to its location.

Pre-job Equipment Opening Meeting or Workshop:

- Ensure that a pre-job equipment opening meeting or workshop is held with operations and maintenance craftsmen to ensure everyone understands the scope and work expectations, including all the following points.
  - An agenda should be developed prior to this important meeting and include the discussion of the following as a minimum:
    - Discuss the work scope, what equipment will be opened, and the opening points.
    - Review the opening process equipment (OPE) tags and their role in managing work and equipment opening. Emphasize that these tags ensure the equipment that remains pressurized is not accidentally opened.
    - All OPE tags should be developed specifically for OPE use, specially designed, weather-resistant, and properly attached directly to the flange or access plate to be opened.
    - Review any work restrictions, including opening any flanges or access plates without an "Open Here" tag attached.
    - Discuss how zero energy will be accomplished and verified.

- Identify the factors that should lead to the immediate shutdown of all work activities. For example: a vapor release in the area, an injury on the job, a leak, a spill, or other containment loss of product, a fire in the immediate vicinity, etc.
- Discuss any required personal protective equipment (PPE) or specific precautions applicable to the work.
- This is a rare chance to work directly with those doing the work; therefore, discuss any other topics that might be unique to either the process, the unit involved, or the work to be done.

Key Equipment and Piping Isolation Points (Key Isolation Points):

- Identify the key isolation points that will be used to isolate the equipment and piping for the opening of process equipment:
  - Create a list of all "key isolation points" used to disconnect or open equipment and piping circuits, including piping isolation for blind installation or removal.
  - The key isolation points are the valves used to isolate piping or equipment. These should be a "double-block and bleed" design and positioned as close to the vessel and equipment as possible.
    - If double block and bleed valves are not available, additional mitigation may be required, especially if the isolation valve is leaking internally.
  - Develop key isolation point tags to identify all the key isolation points for each piping circuit or equipment to be opened.
    - All OPE key isolation point tags should be developed specifically for OPE use, specially designed, weather-resistant, and properly attached directly to the key isolation point (typically an isolation valve) after the valve is closed or properly positioned.
  - The key isolation points can be combined with energy prevention chains and locks to meet the Energy Lockout/Tagout (LOTO) requirements.

Zero-energy Verification Points:

- Identify the "zero-energy verification points," the key vents and drains used to verify zero energy in each piping segment, and all equipment associated with the equipment and piping circuit to be opened?
  - Develop zero-energy verification point identification tags to identify the vents, drains, and bleeders used as zero-energy verification points.
    - All zero-energy verification point identification tags should be developed specifically for OPE use, specially designed, weather-resistant, and properly attached directly to the vent, drain, or bleeder to be opened.

    - Ensure that each energy verification point vent or drain has been verified as unobstructed or internally plugged, including using a bleeder unplugging device if necessary.
    - Leakage from a zero-energy verification vent or drain valve indicates that zero energy hasn't been reached, and further mitigation will be necessary.

Equipment Opening "Open Here" Tags:

- Develop distinctive equipment opening "Open Here" tags and ensure their use is communicated to all craftsmen working on this project.
    - The "Open Here" equipment identification tags should have a distinctive appearance and be designed to stand out when placed on equipment in the field. These distinctive tags should not be used for any other purpose.
    - The distinctive "Open Here" tags will be placed on each piping flange or vessel access plate after the equipment is prepared for opening.
        - All "Open Here" OPE tags should be developed specifically for OPE use, specially designed, weather-resistant, and properly attached directly to the flange or access plate to be opened.
    - All craftsmen should be instructed that they shall not open any equipment or piping not identified with these distinctive "Open Here" tags.

Joint Worksite Visit and Walkdown:

- When issuing the work permit for equipment opening, including blind installation or removal, conduct a "joint worksite visit and walkdown" with the work permit issuer(s) and the workers.
    - Discuss how the equipment has been prepared, including how it has been drained and confirmed as free of hydrocarbons, chemicals, and pressure.
        - Point out the equipment key isolation points and the energy verification points.
        - Point out the "Open Here" tags and how they are used only on the flanges or access plates to be opened.
    - This joint worksite visit and walkdown should be conducted in the field, and all job sites must be visited.
    - The expected job scope should be discussed, including the use of the OPE identification tags and how they will be utilized on this job. Remind the craftsmen that they are not to open any flange or access plate unless it has an "Open Here" tag attached directly to it.
    - Each flange and access plate scheduled for opening should be inspected, and the equipment preparation, including how zero energy has been verified, should be discussed.

  - The use of OPE tags and how they are used to control equipment opening should be discussed during the joint worksite visit and walkdown.
  - Discuss the use of appropriate personal protective equipment, fire protection, emergency egress, and conditions that should lead to stopping work.

Cautionary note: As work begins and continues, it is very important that the operations team revisit the site at regular intervals to ensure that the job is fully compliant with all requirements of the work permit and all other safety regulations for the facility.

It is especially noteworthy that if the job is stopped for a period of time, possibly more than an hour, the equipment and worksite should be re-gas tested to guarantee that it remains within safe limits for entry or hot work.

Blind Removal and Return to Service:

- A similar process to what is outlined here should be used when issuing work permits for blind removal.
  - It is especially important that zero energy is reconfirmed on both sides of installed blinds before issuing work permits for blind removal.
  - This is important because loss-of-containment incidents have also occurred when workers were removing blinds. In each case, process pressure had built up on the isolated side of the blind and went undetected.
  - All vents and drains previously identified with energy verification tags should be closed before admitting hydrocarbons or other chemicals into the unit.

Additional References

American Institute of Chemical Engineers, Line Opening – Strategies & Effective Practices to Manage and Mitigate Hazards, https://www.aiche.org/ccps/resources/tools/safe-work-practices/line-opening/strategies-effective-practices-manage-and-mitigate-hazards.

Blanchard Refining Company LLC – Galveston Bay Refinery, PR-11 Safe Equipment Preparation, https://www.marathonrefinerycontractor.com/content/documents/Refinery_Contractor/Galveston_Bay/SafeProceduresandForms/Safe%20Equipment%20Preparation%20Guidelines.pdf.

ExxonMobil, Opening Process Equipment – Important Considerations, https://refiningcommunity.com/wp-content/uploads/2017/07/Opening-Process-Equipment-OPE-Important-Considerations-Moloney-ExxonMobil-DCU-FCCU-SRU-Budapest-2017.pdf.

Mosaic Chemical Company, Line Break & Equipment Opening Program, https://contractorsweb.mosaicco.com/pdf/EHS_Phos_Program___Line_Breaking_and_Equipment_Opening.pdf.

US Occupational Safety and Health Administration (OSHA), 29 CFR 1910.147, The Control of Hazardous Energy (Lockout/Tagout), https://www.osha.gov/laws-regs/regulations/standardnumber/1910/1910.147.

Westlake Chemical Company, Safe Process Piping, Equipment & Vessel Opening Procedure, https://www.westlake.com/sites/default/files/Safe%20Process%20Piping,%20Equipment%20and%20Vessel%20Opening%20Procedure.pdf.

Index

Note: Page numbers in *italics* and **bold** refers to figures and tables respectively.

Printed and bound by CPI Group (UK) Ltd, Croydon, CR0 4YY

17/09/2026

14964236-0001